ACS SYMPOSIUM SERIES **495**

Catalytic Control of Air Pollution
Mobile and Stationary Sources

Ronald G. Silver, EDITOR
Allied-Signal Automotive Catalyst Company

John E. Sawyer, EDITOR
Allied-Signal Automotive Catalyst Company

Jerry C. Summers, EDITOR
Allied-Signal Automotive Catalyst Company

Developed from a symposium sponsored
by the Division of Colloid and Surface Chemistry
of the American Chemical Society
at the Fourth Chemical Congress of North America
(202nd National Meeting of the American Chemical Society),
New York, New York,
August 25–30, 1991

American Chemical Society, Washington, DC 1992

Library of Congress Cataloging-in-Publication Data

Catalytic control of air pollution: mobile and stationary sources /
 Ronald G. Silver, editor, John E. Sawyer, editor, Jerry C. Summers,
 editor.

   p.   cm.—(ACS symposium series, ISSN 0097–6156; 495)

"Developed from a symposium sponsored by the Division of Colloid
and Surface Chemistry of the American Chemical Society at the Fourth
Chemical Congress of North America (202nd National Meeting of the
American Chemical Society), New York, New York, August 25–30,
1991."

Includes bibliographical references and indexes.

ISBN 0–8412–2455–2

   1. Air—Purification—Equipment and supplies—Congresses. 2. Automobiles—Catalytic converters—Congresses. 3. Catalysis—Congresses.

   I. Silver, Ronald G., 1962–   . II. Sawyer, John E., 1953–   .
III. Summers, Jerry C., 1942–   . IV. American Chemical Society.
Division of Colloid and Surface Chemistry. V. Chemical Congress of
North America (4th: 1991: New York, N.Y.) VI. American Chemical
Society. Meeting (202nd: 1991: New York, N.Y.) VII. Series.

TD889.C37   1992
628.5'3—dc20   92–15615
                                                              CIP

The paper used in this publication meets the minimum requirements of American National
Standard for Information Sciences—Permanence of Paper for Printed Library Materials, ANSI
Z39.48–1984.   ∞

Copyright © 1992

American Chemical Society

ACS Symposium Series

TD889 C37 1992 CHEM

M. Joan Comstock, *Series Editor*

1992 ACS Books Advisory Board

Foreword

THE ACS SYMPOSIUM SERIES was founded in 1974 to provide a medium for publishing symposia quickly in book form. The format of the Series parallels that of the continuing ADVANCES IN CHEMISTRY SERIES except that, in order to save time, the papers are not typeset, but are reproduced as they are submitted by the authors in camera-ready form. Papers are reviewed under the supervision of the editors with the assistance of the Advisory Board and are selected to maintain the integrity of the symposia. Both reviews and reports of research are acceptable, because symposia may embrace both types of presentation. However, verbatim reproductions of previously published papers are not accepted.

Contents

Preface

IMPROVING THE QUALITY OF OUR ENVIRONMENT has become a growing concern in this country and around the globe. Limiting the amount of pollution released into the atmosphere is an important part of that effort. Some of the most promising controls for air pollution involve the use of catalysts. Catalysts are used to control emissions from both mobile sources, such as automobiles, and stationary sources, such as industrial plants. In order to improve and expand the capabilities of these pollution controls, it is important to understand the catalyst chemistry of these systems.

Research efforts in this field have recently been accelerated by the passage of the 1990 Clean Air Act. This legislation tightens the limits on currently controlled emissions and expands the list of regulated emissions. The new law also introduces the concept of alternative fuels for lower emissions from vehicles. New laws in California will also impose new standards and will require improved technologies to meet them. However, air pollution control is not limited to the United States but is in fact a worldwide concern.

This volume is based on a symposium that is part of a continuing series on the surface science of catalysis. The symposium was organized to provide a forum for recent developments in the area of air pollution control via catalysis. Catalysis for the areas of both mobile source pollution control and stationary source pollution control were explored. Authors from industry and academia were included in each area.

Including stationary and mobile source chapters alike in one volume allows the reader to note the similarities and differences between the two fields and possibly to apply ideas from one area to the other. The coverage is not intended to be exhaustive but rather to serve as a survey of some of the most current topics of interest in this field. The intended audience for this book is the chemist or engineer interested in pollution control, or prevention, or both in the automotive, chemical, petroleum, and other industries, or otherwise involved in the environmental applications of catalysts.

The book is split into two sections. The first deals with mobile source catalyst pollution control, and the second with stationary source control. Even though some important differences exist between the two areas, the catalyst chemistry involved is the same or similar and justifies including both areas in one book.

The first section includes an overview of recent mobile source emission control legislation, discussions of catalysts for alternative fuels such as natural gas or methanol, a comparison of hydrocarbon reactivities, and studies of the effects that surface properties of catalysts have on reactions for pollution control. The second section includes an overview of the regulations for stationary sources and examples of catalytic reduction, oxidation, combustion, and thermal decomposition of pollutants.

RONALD G. SILVER
JOHN E. SAWYER
JERRY C. SUMMERS
Allied-Signal Automotive Catalyst Company
Tulsa, OK 74158

January 3, 1992

Mobile Source Emission Control

Chapter 1

Catalyst Technologies To Meet Future Emission Requirements for Light-Duty Vehicles

Jerry C. Summers and Ronald G. Silver

Allied-Signal Automotive Catalyst Company, P.O. Box 580970, Tulsa, OK 74158-0970

New automotive catalyst technologies are being developed for vehicle emission systems designed to meet new Federal emissions standards effective in 1994. The new standards feature a 30% reduction in non-methane HC emissions, a 60% reduction of NOx and a 100% increase (to 100,000 miles for light duty vehicles) in warranty period during which these vehicles must meet emission standards (called the "useful life" requirement). In addition, California has developed its own ambitious requirements for significant future reductions in emissions.

California's emission reduction requirements introduce a number of new terms into our vocabulary including transitional low emission vehicle (TLEV) (HC/CO/NOx = 0.125/ 3.4/ 0.4g/mi), low emission vehicle (LEV) (0.075/3.4/0.2g/mi), ultra-low emission vehicle (ULEV) (0.04/1.7/0.2g/mi) and zero emission vehicle (ZEV) (1). With each level of increasingly stringent emission reduction comes an array of technologies and emission control strategies envisioned as nesessary for achieving the designated emission targets. HC and NOx emission are the main targets of the California program. These gaseous pollutants react in the presence of sunlight to form ozone, the most pervasive air pollution problem facing the state.

New automotive emission control standards are being legislated in other parts of the world as well. In Europe, starting in July of 1992, the EEC will limit combined emissions of HC and NOx to 0.97 g/km (1.55g/mi), CO will be limited to 2.72 g/km (4.35 g/mi), and particulate matter will be limited to 0.14 g/km (0.22 g/mi), for both gasoline and diesel powered passenger cars. Tighter emission standards will be introduced starting in 1996, pending approval of the EEC Council of Ministers. In other areas of Europe, the Stockholm Group currently has the same standards as the United States, but is expected to tighten its NOx standard by about 50% by 1995. For diesel cars, the Stockholm group will require a particulate standard of 0.08 g/km (0.13g/mi).

In East Asia, Japan has had emission controls as long as the United States. Since 1978, the Japanese have been meeting a standard of 0.25 g/km (0.4 g/mi) for HC, 2.1 g/km (3.4 g/mi) for CO and 0.25 g/km (0.4 g/mi) for NOx. Except for the HC standard, which is slightly less severe for Japan, these are the same as the new American standards. South Korea has recently introduced standards, for cars with engines larger than 800cc, of 8.0 g/km

0097–6156/92/0495–0002$06.00/0

(12.8 g/mi) for CO, 1.5 g/km (2.4 g/mi) for NOx, and 2.1 g/km (3.4 g/mi) for HC. By the end of the decade, emission controls will also be required for engines smaller than 800cc. Lower cost catalysts are also a priority with this emerging third world nation.

Automobiles are not the only emission sources being considered for controls. Regulations are also being formulated for motorcycles in some countries. For example, by 1994 in Taiwan motorcycle emission standards will be 1.2 g/km (1.9 g/mi) for HC, 1.5 g/km (2.4 g/mi) for CO and 0.4 g/km (0.64 g/mi) for NOx. Future standards for motorbikes in Europe may be modelled on the current Swiss standards of 1.0 g/km (1.6 g/mi) HC and 1.2 g/km (1.92 g/mi) for CO.

Controlling Hydrocarbon Emissions

The HC control requirements adopted by California necessitate considerable emission control improvement during the initial phase of the Federal Test procedure (FTP-75 cycle). Roughly 60-80% of the engine-out HC emissions for a typical vehicle occur during this phase (called the "Bag 1 or cold start phase). Major reductions in Bag 1 emissions will result in major reductions of overall vehicle emissions.

Various technology scenarios have been developed which correspond to the increasingly stringent legislated HC standards in California. Table I shows these standards and the technologies proposed to meet these standards.

The first tier, called transitional low emission vehicle standards, calls for a non-methane hydrocarbon standard of 0.125 g/mile. This may be achieved by placing a catalyst close to the engine manifold (close-coupled) on small conventional gasoline vehicles. The use of reformulated gasolines or alternative fuels, combined with the appropriate catalyst technology, may also meet this hydrocarbon standard.

The second tier standard, called low emission vehicle standards, will require a non-methane hydrocarbon limit of 0.075 g/mile. The technologies proposed to meet this challenge include the addition of electrically heated catalysts (2,3) to conventional gasoline vehicles, where the catalyst could be heated to its most efficient operating temperature before the vehicle starts, and alternate-fueled vehicles (such as CNG or LPG fuel) combined with close-coupled catalysts.

The next tier, called ultra-low emission vehicle standards, will limit non-methane hydrocarbon emissions to 0.04 g/mile. This standard may possibly be achieved by adding an electrically heated catalyst to either a small conventional gasoline vehicle, or a reformulated gasoline or methanol fueled vehicle. Other ways of meeting this challenge include combining a natural gas vehicle with a close-coupled or heated catalyst, or combining a propane fueled car with a heated catalyst. Finally, the ultimate standard of zero emissions will probably only be attainable by electric vehicles.

The above standards will be gradually phased in from 1994 to 2003, as indicated in Table II. The percentages shown in the table represent the percentage of new cars sold in California required to meet the indicated standard.

In recent years, much interest has been expressed in using clean fuels as replacements for conventional gasolines, in order to meet the challenge imposed by these proposed California standards. The California Air Resources Board has formulated a plan to identify useful technologies and to arrange demonstrations of their usefulness for achieving a considerable reduction of reactive hydrocarbon emissions from mobile sources. The use of methanol or compressed natural gas as alternative fuels is receiving high

Table I. Projections of Available Technology to Meet Proposed Standards

Tier Abbreviation	Conventional Gasoline Based Standards (g/mi) @ 50K Miles			Projected Technology Able to Meet Standards
	NMHC	CO	NOx	
TLEV	0.125	3.4	0.4	- Small Conventional Gasoline Vehicles with Close-Coupled Catalysts. - Reformulated Gasoline/M85 Vehicles with Catalysts. - CNG/LPG Vehicles with Catalysts. - Hybrid electric/Electric Vehicles
LEV	0.075	3.4	0.2	- Conventional Gasoline Vehicles with Electrically-Heated Catalysts. - CNG/LPG Vehicles with Close-Coupled Catalysts. - Hybrid electric/Electric Vehicles
ULEV	0.040	1.7	0.2	- Small Conventional Gasoline Vehicles with Electrically-Heated Catalysts. - Reformulated Gasoline/M85 Vehicles with Electrically-Heated Catalysts. - CNG Vehicles with Close-Coupled or Electrically-Heated Catalysts or LPG Vehicles with Electrically-Heated Catalysts. - Hybrid electric/Electric Vehicles.
ZEV	0	0	0	- Electric Vehicles.

Table II. California Non-methane Hydrocarbon Implementation Strategy

Model Year	.39g/mi	.25g/mi	TLEV .125g/mi	LEV .075g/mi	ULEV .04g/mi	ZEV .00g/mi	Average (g/mi)
1994	10%	80%	10%				0.252
1995		85%	15%				0.231
1996		80%	20%				0.225
1997		73%		25%	2%		0.202
1998		48%		48%	2%	2%	0.157
1999		23%		73%	2%	2%	0.113
2000				96%	2%	2%	0.073
2001				90%	5%	5%	0.070
2002				85%	10%	5%	0.068
2003				75%	15%	10%	0.062

priority. Both fuels have much lower non-methane HC emissions relative to current gasoline. As a less costly alternative to fuels such as these, the major petroleum companies are re-formulating the compositions of their gasolines with an eye towards reducing the levels of toxic and reactive hydrocarbons emitted during vehicle service. This includes not only tailpipe emissions, but also evaporative emissions and refueling losses. Several reformulated gasoline formulations have already been commercially introduced.

Catalytic Approaches for Meeting the HC Control Challenge

Three general approaches to improve Bag 1 HC control appear promising: the use of electrically-heated metallic supported catalysts (2,3), close-coupled conventional catalysts, and moving underfloor catalysts closer to the engine. One "secret" of good catalytic HC control is a rapid heat up of the catalyst.

The first approach, using an electrically-heated metallic substrate housed near the engine manifold, has attracted considerable attention. This technology is under aggressive development by the major metallic substrate companies. The technology consists of a catalyzed metallic foil substrate, a power source and control logic and hardware.

In operation, the metallic substrate is first heated by electric current for a period of time (the shorter the better) prior to vehicle start up. A short warm-up time will minimize the inconvenience to the driver waiting to put his vehicle in motion. Current metallic substrates are heated for about 15 seconds before the engine is started, and then for an additional 30 seconds after starting the engine in order to heat the cold exhaust gases (4). Development work is underway to minimize the time needed for warm-up prior to engine ignition. A recent study claims to have achieved light-off over a catalyst on a metallic substrate within 20 seconds, without preheating the substrate prior to starting the engine (5). Improving the heat capacity or heat transfer characteristics of the monolith substrate will also reduce the time required to heat the catalyst to achieve light-off.

Simultaneously with heating the catalyst, small quantities of air are added to the exhaust so that its Air-to-Fuel (A/F) is net lean - an A/F condition that increases the ease in which catalysts oxidize both HC and CO emissions. Low mileage emissions tests with electrically-heated catalyst equipped vehicles have resulted in HC emission levels at or below the ULEV HC standard (4). While this technology shows promising results, further development is necessary. Among other things, an adequate power source and electronic control system need to be identified and the durability of these components need to be demonstrated.

A second approach for improved HC control is to add a non-resistively-heated monolith catalyst (either ceramic or metallic substrate) off the engine manifold. For this approach and for the previous one, it is probable that underfloor converters will also be required to meet the California low and ultra-low emission standards.

The third approach, moving an underfloor catalyst closer to the engine manifold, will also help improve HC control. As a rough rule of thumb, the exhaust temperature increases about 55 °C for each foot the catalyst is moved closer to the manifold. While hotter catalyst inlet temperatures favor improved HC and CO conversion performance, NOx conversion increases with increasing temperature up to about 400°C. Somewhat above that temperature, it begins to decline. Thus, there are performance trade-offs to be considered with increasing exhaust temperature when locating the catalytic converter closer to the manifold. Care must be exercised to optimize catalyst inlet temperature to achieve the desired levels of HC/CO/NOx performance.

Higher operating temperatures, coupled with an extended useful life period, require a new generation of high activity/thermally stable catalysts. New thermally stable washcoats, which contain the active noble metals, are being developed to withstand increasingly severe operating temperatures. These washcoats must minimize sintering and alloy formation of the various noble metals. The composition of the washcoat should be a function of the noble metal composition. A variety of base metals are used to stabilize the washcoat alumina against sintering as well as promoting noble metal activity and stabilizing noble metal dispersions.

In the case of methanol fueled vehicles, formaldehyde emissions present a special concern. Catalysts must be formulated which oxidize formaldehyde to CO_2 and H_2O. Formaldehyde is especially undesirable due to its high photochemical reactivity (6). It is also a lacrimator and a known carcinogen. A formaldehyde certification standard of 15 mg/mile and an in-use standard of 25 mg/mile for 1993 - 1995 has been enacted by the California Air Resources Board (CARB) for methanol fueled vehicles. Beginning in 1996, methanol vehicles will be required to meet an in-use standard of 15 mg/mile for formaldehyde emissions, and a standard of 8 mg/mile is under consideration for the late 1990s (Bertelson, B., Manufacturers of Emission Controls Association, personal communication, 1991).

Vehicles running on methanol tend to run at lower temperatures than conventional gasoline cars. The low temperatures in turn, lead to problems with catalyst lightoff, and thus to low conversions during Bag 1. Vehicles may also be difficult to start under cold temperatures, due to the low volatility of the fuel. Some catalysts may be placed near the manifold for aldehyde control, so thermally stable washcoats will be needed for high temperatures in this application.

Variable fueled vehicles are now being introduced which have the ability to run on gasoline, methanol, or a gasoline/methanol blends. Catalysts designed for these vehicles will also have to perform acceptably when the car is fueled with either gasoline or methanol.

Hydrocarbon emissions from natural gas fueled vehicles may also pose a challenge. The 1990 Clean Air Act introduced a standard of 0.25 g/mile for "non-methane organic gases". It did not, however, specifically repeal the old standard of 0.41 g/mile for total hydrocarbons. Therefore, the courts may have to rule on whether vehicles will be required to meet both standards, or only the non-methane standard. If both standards must be met, then the methane emissions from the natural gas vehicles will need to be reduced. This presents a significant challenge to the catalyst manufacturer, since methane is extremely resistant to oxidation (7). Bag 1 emissions will be especially difficult to reduce, since methane lightoff occurs at relatively high temperatures (500 - 600 °C) which conventional underbody catalysts do not reach until well into the emissions test.

The air/fuel mixture entering a CNG fueled engine will require a feedback control system specially designed for natural gas, since adequate control of the air/fuel mixture with a system built for gasoline is difficult to achieve (8). Another area of concern is with the high temperature durability of the natural gas catalyst, especially since to effectively convert methane the catalysts will probably have to be installed near the engine manifold where they will be exposed to high temperatures (550 °C or higher) during in-service operation.

Controlling NOx Emissions

As discussed previously, improvements in HC and CO control are best achieved by rapidly heating up the catalyst in Bag 1 to obtain acceptable performance. In contrast, NOx emissions become a more serious problem during higher temperature operation. Thus, the catalytic control strategy for NOx control is somewhat different than for HC control. The Federal government recently tightened the NOx standard from 1.0 to 0.4 g/mile, while California will tighten theirs from 0.4 to 0.2 g/mile beginning in 1997 (ULEV standard). For catalytic NOx control, the use of certain base metals in washcoat formulations are of paramount importance in promoting the activity and durability performance of the noble metals. Cerium oxide is perhaps the most commonly used base metal oxide for these purposes (9-11). Stabilizing rhodium to improve NOx thermal durability performance, has been, and will continue to be a priority in developing improved NOx control catalysts.

Currently the car companies' primary emission engineering objectives are to develop systems that meet the new emission standards. Once this is accomplished, they will undoubtedly focus more of their efforts on cost containment. They may well include smaller catalyst volumes, less noble metal usage and the substitution of less expensive noble metals for the more expensive ones. Palladium may be used increasingly for close-coupled applications since it has superior thermal resistance relative to platinum and rhodium. Also, new palladium-only formulations show considerable promise for not only improved light-off performance, but also for warmed-up three-way performance (12-14). It is likely that considerable effort will be expended to keep the noble metal cost as low as possible without adversely affecting the emissions systems ability to meet the standards.

Palladium catalysts have long been recognized as having desirable performance properties. In addition to its ready availability and low cost relative to platinum and rhodium, it is superior to platinum for CO oxidation and oxidation of unsaturated hydrocarbons (15). However, palladium is more susceptible than platinum to tetraethyl lead poisoning (16), and is also susceptible to poisoning by sulfur oxides (17,18).

When emission standards have been tightened in the past, it has been common practice to increase noble metal content, catalyst volume or the number of catalytic units per vehicle. Surely we can expect to see examples of the practice continued through the 1990s.

There is a growing pressure on car producers to make significant fuel economy gains. Accomplishing this task would also cut the level of emissions of the important greenhouse gas carbon dioxide as well as promoting energy security. One strategy for achieving this objective requires that a vehicle operate significantly lean of the stoichiometric point (A/F = 18-22). Figure 1 shows the affect of air/fuel ratio on emissions from a natural gas vehicle. For gasoline, the stoichiometric point is around A/F = 14.56, and for natural gas the stoichiometric point is around A/F = 16.5. Operating under fuel-lean conditions results in low CO and HC emissions. Even though NOx emissions are low, further reduction is required to meet the current emission standards, as illustrated in Figure 1. Conceivably, this might be accomplished via the NO-CO reaction, by direct NO decomposition to N_2 and O_2, or by reacting NO with hydrocarbon in the presence of oxygen. The most promising process appears to be the reaction of NO with hydrocarbon. Thus, a new type of automotive catalyst technology, referred to as lean NOx catalysis, is under active development. The key feature of this process appears to be the reaction of NOx with feedstream HC in the presence of oxygen to yield nitrogen. The rate and extent of this reaction is a strong function of the available HC.

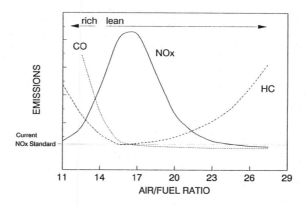

Figure 1. Effect of A/F ratio on emissions from a natural gas fueled engine.

The catalyst formulation that has received the most attention to date is copper incorporated into the zeolite ZSM-5 (19,20). Other base metals (Co, Ni, Fe, Mn, Cr, for example) and other support materials (such as other zeolites, mordenite, spinels and perovskites) are also actively being researched for potential lean NOx control.

Recent announcements have indicated that lean NOx control catalysts are virtually ready for commercialization (21). Unfortunately, that is not the case. Further development work on this promising technology is needed. While some lean NOx control catalysts give excellent fresh NOx performance (60-80% conversion) at A/F = 18-22, durability is a significant issue to be addressed and adequate levels of performance have yet to be achieved.

Catalyst Technology Will Play a Key Role in Meeting Future Cleanup Challenge

Despite significant progress in cleaning up the air we breathe, much work remains. The new Federal and California vehicle emission standards will help us move towards cleaner air. These tougher standards, however, offer significant engineering challenges. Catalyst technology, which has played a key role in significantly reducing emissions from motor vehicles over the past 15 years, will continue to advance and will help meet the emission clean-up challenges of the 1990s and beyond.

Catalysts for automotive emission control make up a significant part of the world-wide catalyst industry. Out of the $5 billion market for catalysts world-wide in 1989 (22), approximately $1.5 billion were spent for emission control catalysts. This year, about 45 million catalyst monoliths are expected to be produced, for a market valued at close to $2 billion. By 1996, those numbers may rise to 67 million catalyst pieces and a $3.5 billion market. Of course, new and improved technology will play an important role in this rapidly expanding market, and the chemistry behind this technology will be discussed in more detail in the chapters which follow in this section.

Acknowledgments

The authors would like to thank W. B. Williamson for his technical consultation, P. M. Caouette for typing the manuscript, and R. Hefferle for the marketing estimates.

Literature Cited

1. Staff Report, "Proposed Regulations for Low-Emission Vehicles and Clean Fuels," prepared by the State of California Air Resources Board, Mobile Source and Stationary Source Divisions, Release Date: August 13, 1990.
2. Hellman, K. H., Bruetsch, R. I., Piotrowski, G. K., and Tallent, W. D., *SAE Paper No. 890799* (1989).
3. Whittenberger, W. A., and Kubsh, J. E., *SAE Paper No. 900503* (1990).
4. Heimrich, M.J., Albu, S. and Osborn, J., *SAE Paper 910612*, (1991).
5. Gottberg, I., Rydquist, J. E., Backlund, O., Wallman, S., Maus, W., Bruck, R. and Swars, H., *SAE Paper No. 910840* (1991)
6. Lowi, A. and Carter, W. P. L., *SAE Paper No. 900710* (1990).
7. Y-F. Y. Yao, *Ind. Eng. Chem. Prod. Res. Dev. 19* , (1980), p.293.
8. Summers, J. C., Frost, A. C. , Williamson, W. B., and Freidel, I. M., "Control of NOx/CO/HC Emissions from Natural Gas Fueled Stationary Engines with Three-Way Catalysts", presented at the 84th Annual Meeting

of the Air & Waste Management Association, Vancouver, B. C., June 16-21, 1991.
9. Hegedus, L. L., Summers, J. C., Schlatter, J. C., and Baron, K., *J. Catal. 54,* (1979), p. 321.
10. Kim, G., *Ind. Eng. Chem. Prod. Res. Dev. 21,* (1982), p.267.
11. Cooper, B. J., and Truex, T. J., *SAE Paper No. 850128* (1985).
12. Muraki, H., Shinjoh, H., Sabukowa, H., Yokata, K., and Fujitani, Y.,*Ind. Eng. Chem. Prod. Res. Dev. 25,* (1986), p. 202..
13. Summers, J. C., Williamson, W. B., and Henk, M. G., *SAE Paper No. 880281* (1988).
14. Summers, J. C., White, J. J., and Williamson, W. B., *SAE Paper No. 89074* (1989).
15. Barnes, G. J., and Klimisch, R. L., *SAE Paper No. 730570* (1973).
16. Klimisch, R. L., Summers, J. C., and Schlatter, J. C., in *Catalysts for the Control of Automotive Pollutants* , McEvoy, J. E., Ed., Adv. Chem. Series No. 143, ACS Books: Washington, DC, 1975, pp.103-115.
17. Monroe, D. R., Krueger, M. H., and Upton, D. J., "The Effect of Sulfur on Three-Way Catalysts", 2nd International Congress on Catalysis and Automotive Pollution Control (CAPoC 2) Brussels, Sept. 1990.
18. Sims, G., *SAE Paper 912639,* (1991).
19. Iwamoto, M., Yahiro, H., and Tanda, K., in *Successful Design of Catalysts* , Elsevier: Amsterdam, 1988, pp. 219-226.
20. Kummer, J. T., "Catalysts for NO Decomposition Under Oxygen Rich Conditions", in *Proc. of the Gas Rsrch. Inst. Nat. Gas Eng. Catal. Wkshp.,* (March 1991), GRI-91/0197.
21. "A Zeolite Catalyst for Lean-Burning Engines",*Chem. Engr.*, (April 1991), p.19.
22. Greek, B. F, *Chem. and Eng. News 67* , No. 22, (May 29, 1989) , p.29.

RECEIVED April 3, 1992

Chapter 2

Methane Oxidation over Noble Metal Catalysts as Related to Controlling Natural Gas Vehicle Exhaust Emissions

S. H. Oh[1], P. J. Mitchell[1], and R. M. Siewert[2]

[1]Physical Chemistry Department and [2]Thermosciences Department, General Motors Research Laboratories, Warren, MI 48090

Natural gas has considerable potential as an alternative automotive fuel. Methane, the principal hydrocarbon species in natural-gas engine exhaust, has extremely low photochemical reactivity but is a powerful greenhouse gas. Therefore, exhaust emissions of unburned methane from natural-gas vehicles are of particular concern. This laboratory reactor study evaluates noble metal catalysts for their potential in the catalytic removal of methane from natural-gas vehicle exhaust. Temperature run-up experiments show that the methane oxidation activity decreases in the order Pd/Al_2O_3 > Rh/Al_2O_3 > Pt/Al_2O_3. Also, for all the noble metal catalysts studied, methane conversion can be maximized by controlling the O_2 concentration of the feedstream at a point somewhat rich (reducing) of stoichiometry.

In recent years natural gas has received increased attention as an alternative fuel for motor vehicles because of its potential technical, economic and environmental advantages. Natural gas, which consists primarily of methane (85–95% of the total HC), has excellent knock resistance and ignition capability over a wide range of air-fuel ratios (1). These properties permit natural-gas engines to operate at high compression ratios and with very fuel-lean mixtures, resulting in substantially higher fuel efficiencies than are possible with gasoline engines. Also, natural gas is less expensive than gasoline on an energy basis and is readily available from abundant domestic supplies (2, 3). Environmental benefits of natural gas include extremely low photochemical reactivity (4), reduced cold-start CO emissions (1), and zero evaporative emissions. Furthermore, the low carbon content of methane would lead to reduced CO_2 emissions (the most common "greenhouse" gas) from natural gas vehicles. However, methane itself is a much more powerful greenhouse gas than CO_2 (3, 5) and thus, though not currently regulated, exhaust emissions of unburned methane from natural-gas vehicles are anticipated to be of particular concern in the future.

Vehicle emission tests conducted at General Motors Research Laboratories with two different gasoline engines converted to operate on natural gas (6) showed that poor methane conversion (<15% during FTP) occurs over commercial three-way catalysts currently used for gasoline-vehicle exhaust emission control. These dual-fuel vehicles were designed to operate open-loop (lean) when running on compressed natural gas. This observation of low methane conversion in the lean natural-gas engine exhaust is not surprising in view of previous reports in the literature (Ref. (7) and references therein), which generally show that methane is the most difficult hydrocarbon to oxidize catalytically. The results of these vehicle emission tests also indicate that the optimal catalyst formulation for natural-gas vehicles may differ from that for gasoline vehicles.

This laboratory reactor study was initiated to investigate the possibility of controlling methane emissions by catalytic oxidation of unburned methane present in natural-gas engine exhaust. Most of the previous laboratory studies on the catalytic oxidation of methane [e.g., Refs. (7-11)] focused on experimental conditions where there is an excess of oxygen over methane (i.e., $O_2/CH_4 > 2$). Because of the wide ignition range of natural gas, however, actual natural gas engines can potentially be operated under a variety of air-fuel ratio conditions. The currently favored operating strategies for natural gas engines include fuel-lean and stoichiometric combustion (1). Since catalytic activity is often a sensitive function of the stoichiometry of the reaction environment (12), it is entirely possible that methane conversion efficiency in natural-gas engine exhaust may vary substantially with the air-fuel ratio. With this possibility in mind, we conducted laboratory methane oxidation experiments with single-component noble metal catalysts over wide ranges of temperatures and feedstream stoichiometries. The results of such laboratory experiments are of practical interest because (1) methane is the principal hydrocarbon species in natural-gas engine exhaust, (2) its oxidation characteristics have not been examined under conditions likely to be encountered in natural-gas vehicle exhaust, and (3) noble metals are ranked among the most active catalysts for methane oxidation (7,8).

Experimental

Catalysts. All of the catalysts were prepared by incipient wetness impregnation of a γ-alumina support (3 mm diameter spheres, 96 m^2/g BET area) with aqueous solutions of the metal salts. The metal loadings, metal salts used in the preparation, and CO/metal and H/metal ratios determined from static chemisorption measurements are listed in Table 1. The noble metal loadings of the catalysts were chosen to provide a similar number of active metal atoms in all cases. After impregnation the catalysts were dried in air overnight at room temperature and then calcined in flowing air at 500°C for 4 h. Such procedures resulted in the deposition of the noble metals near the periphery of the catalyst beads. The observation of a CO/Rh ratio in excess of unity for Rh/Al_2O_3 indicates that at least part of the Rh exists as isolated atoms or ions, forming a dicarbonyl species during CO chemisorption (13,14).

Table 1. Catalyst Properties

Catalyst	Metal Loading (wt%)	Noble Metal Precursor	Chemisorption	
			CO/M	H/M
Pt/Al_2O_3	0.20 Pt	H_2PtCl_6	0.60	0.70
Pd/Al_2O_3	0.16 Pd	$PdCl_2$	0.21	0.32
Rh/Al_2O_3	0.14 Rh	$RhCl_3$	1.25	0.42

SOURCE: Adapted from ref. 35.

Reactor System and Analytical Methods.   The integral flow reactor
system used in this study was similar to that employed in a previous
study (15).   The reactor was a 2.5 cm o.d. stainless steel tube
housed in an electric furnace.   The feed gases were passed downward
through layers of silicon carbide particles and the catalyst pellets.
The silicon carbide layer located upstream of the catalyst bed serves
as an inert heat transfer medium and also helps establish fully-
developed flow in the reactive section. Temperatures were measured
with a chromel-alumel thermocouple positioned along the reactor
centerline with its tip located a few millimeters below the top of
the catalyst bed.   All the experiments reported here were done using
15 cm^3 of catalyst and a total feedstream flow rate of 13 L/min
(STP), yielding a space velocity of 52 000 h^{-1}. The feedstream
contained 0.2 vol% CH_4, 0.1 vol% CO if present, and variable levels
of O_2 in a He background.
        The gas stream entering and leaving the reactor was analyzed
using a Varian 6000 gas chromatograph equipped with a thermal
conductivity detector. A single column (0.32 cm diameter by 1.5 m
length) containing Molecular Sieve 5A was employed, and the
chromatographic separations were carried out isothermally at 60°C in
a He carrier gas.   Individual species in the reaction mixture were
identified and quantified by comparing their elution times and
integrated areas with those of commercially-supplied calibration
gases.
        The methane oxidation activity of the catalysts was
characterized in two ways: (1) temperature run-up experiments with
fixed feed composition, and (2) variable composition experiments at a
fixed temperature. Since our primary interest here is in evaluating
catalysts (rather than detailed kinetics) for potential automotive
application, we chose to measure catalyst performance under carefully
controlled inlet conditions to the catalyst bed.   All the methane
oxidation activities reported here are those for the fresh catalysts.
The reactor temperature was controlled by a thermocouple placed at
the outside surface of the reactor tube. However, temperature values
quoted in this study (referred to as catalyst temperature) are those
actually measured just below the top of the catalyst bed.

Results

Oxidation Activity of Noble Metal Catalysts in CH_4-CO-O_2 Mixtures. Although our principal focus is on the catalytic oxidation of methane, we first conducted laboratory reactor evaluations of the three noble metals Pt, Pd, and Rh in feedstreams that contained CO in addition to methane and O_2. We added CO to the feed because CO, which is present in natural-gas vehicle exhaust in significant quantities (6), has been shown to significantly affect the oxidation activity of noble metal catalysts for some hydrocarbons (16-18). Furthermore, the presence of CO in the feedstream allows one to examine the conversion efficiencies of both methane and CO simultaneously.

Figure 1 shows the steady-state conversions of CH_4 and CO as a function of temperature for the Pt/Al_2O_3, Pd/Al_2O_3, and Rh/Al_2O_3 catalysts. The experiments were conducted in an oxidizing feedstream containing 0.2 vol% CH_4, 0.1 vol% CO, and 1 vol% O_2. The CO conversion over the Pt/Al_2O_3 and Pd/Al_2O_3 catalysts is near 100% at temperatures as low as ~200°C; however, the Rh/Al_2O_3 catalyst was observed to be much less active for CO oxidation in this oxidizing environment, requiring ~350°C for complete CO conversion. In agreement with earlier literature reports of extremely low methane oxidation rates (10,11,19), high temperatures in excess of 500°C are required for 50% conversion of the methane over the noble metals in the oxidizing feedstream employed in Figure 1. The low reactivity of CH_4 toward O_2 is also reflected in the absence of catalyst lightoff during the methane oxidation. The methane oxidation activity in the presence of excess O_2 decreases in the order Pd/Al_2O_3 > Rh/Al_2O_3 > Pt/Al_2O_3, and this activity ranking is consistent with the observations of Firth and Holland (9) and of Yu Yao (10). In the oxidizing feedstream of Figure 1, carbon dioxide was the only carbon-containing reaction product detected at temperatures above 200°C (where the CO in the feed was completely converted), indicating the complete oxidation of the methane in the feed. Complete oxidation of methane was also observed in CO-free oxidizing feeds, as will be discussed later in this paper.

Since actual natural-gas vehicles can potentially operate over a wide range of air-fuel ratios, it is of practical interest to examine the performance of the noble metal catalysts under net-reducing conditions as well. Figure 2 shows steady-state CH_4 and CO conversions as a function of temperature in a reducing feedstream containing 0.2 vol% CH_4, 0.1 vol% CO, and 0.33 vol% O_2. All three noble metals exhibit very similar low-temperature CO oxidation activity, reaching nearly 100% CO conversion at temperatures as low as 200°C. However, the CO conversion efficiencies over the noble metal catalysts begin to fall below 100% near 450°C and then continue to decline with a further increase in temperature. For each of the noble metal catalysts, the temperature required for the onset of the CH_4 oxidation in the reducing feedstream of Figure 2 is similar to that observed in Figure 1 for the oxidizing feedstream. Notice, however, that once the reaction began, the CH_4 conversion increased much more sharply with temperature (leveling off between 80 and 95% for T > 550°C) in the reducing feedstream than it does in the oxidizing feedstream. The observation that the decrease in CO conversion is generally accompanied by an increase in CH_4 conversion

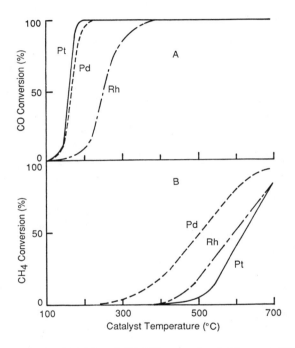

Figure 1. Conversions of (A) CO and (B) CH_4 over the alumina-supported noble metal catalysts as a function of temperature in an oxidizing feedstream containing 0.2 vol% CH_4, 0.1 vol% CO, and 1 vol% O_2. (Reproduced with permission from ref. 35. Copyright 1991 Academic.)

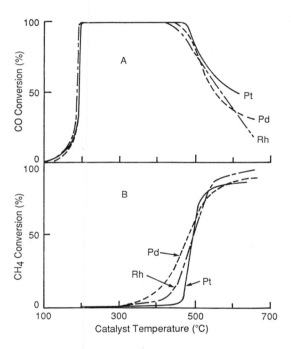

Figure 2. Conversions of (A) CO and (B) CH_4 over the alumina-supported noble metal catalysts as a function of temperature in a reducing feedstream containing 0.2 vol% CH_4, 0.1 vol% CO, and 0.33 vol% O_2. (Reproduced with permission from ref. 35. Copyright 1991 Academic.)

in Figure 2 strongly suggests that the loss in CO conversion efficiency at elevated temperatures is due primarily to partial oxidation of methane to CO under the oxygen-deficient conditions. In fact, our experiments with CH_4-O_2 mixtures (i.e., no CO in the feed) under reducing conditions confirmed the formation of CO as the principal carbon-containing partial oxidation product of methane oxidation. This aspect will be discussed later in the paper.

Comparison of Figures 1B and 2B shows that at temperatures above 550°C the CH_4 conversion over each of the noble metal catalysts is higher in the reducing environment than it is in the oxidizing environment. This suggests the possibility that the CH_4 conversion may be sensitive to the stoichiometry of the gas-phase environment and thus there may exist an optimum feedstream stoichiometry at which a maximum CH_4 conversion occurs. To explore this possibility, we measured steady-state CH_4 and CO conversions at a catalyst temperature of ~550°C using feedstreams containing 0.2 vol% CH_4, 0.1 vol% CO, and variable levels of O_2. A wide range of feedstream stoichiometries were covered by increasing the O_2 concentration step-by-step while holding the temperature and total flow rate constant. The results of such experiments are shown in Figure 3, where the CH_4 and CO conversions are plotted against the O_2 concentration in the feed. It can be seen in Figure 3B that for each of the noble metal catalysts, the CH_4 converison goes through a maximum at an O_2 concentration somewhat less than its stoichiometric value of 0.45 vol% (i.e., in a net-reducing feed) and then declines sharply as the O_2 concentration is increased further. The optimum inlet oxygen concentration for Pt/Al_2O_3 and Rh/Al_2O_3 lies between 0.40 and 0.42 vol% O_2; however, a maximum CH_4 conversion over Pd/Al_2O_3 was observed at a lower inlet O_2 concentration of ~0.35 vol%. The CO conversions measured during the same experiments are plotted in Figure 3A. For all three noble metal catalysts, the CO conversion increases monotonically with increasing O_2 concentration in the feed, reaching an asymptotic level of 100% at an O_2 concentration between 0.45 and 0.5 vol%. Under reducing conditions, the CO conversion efficiency decreases in the order Pt/Al_2O_3 > Pd/Al_2O_3 > Rh/Al_2O_3.

Oxidation Activity of Noble Metal Catalysts in CH_4-O_2 Mixtures. Additional methane oxidation experiments were carried out over the same alumina-supported noble metal catalysts without CO in the feed. Comparison of the results of these experiments with those presented in the previous section allows one to examine how the CH_4 conversion characteristics of the noble metal catalysts are affected by the presence of CO in the feed. Also, experiments under conditions where CO is absent in the reactant gas mixture provide useful insight into the origin of the CO observed during our previous methane oxidation experiments in a reducing environment and the tendencies of the various noble metals to partially oxidize methane to CO.

Temperature run-up experiments conducted in an oxidizing CH_4-O_2 feedstream (0.2 vol% CH_4, 1 vol% O_2, no CO) revealed methane conversion characteristics similar to those depicted in Figure 1B. Similarly, the methane conversion vs. temperature data obtained with a CO-free reducing feedstream containing 0.2 vol% CH_4 and 0.33 vol% O_2 exhibit close similarities to those shown in Figure 2B. The similarities in methane oxidation features under reducing conditions between the cases with and without CO include (1) the onset of

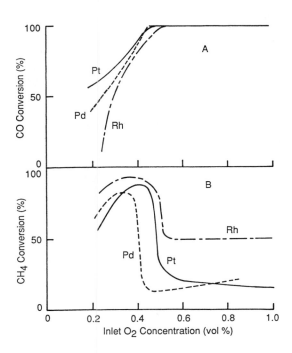

Figure 3. Conversions of (A) CO and (B) CH$_4$ over the alumina-supported noble metal catalysts at a catalyst temperature of 550 °C in feedstreams containing 0.2 vol% CH$_4$, 0.1 vol% CO, and variable levels of O$_2$. (Reproduced with permission from ref. 35. Copyright 1991 Academic.)

methane oxidation at a comparable temperature for each of the noble metals, (2) the same activity ranking (Pd/Al_2O_3 > Rh/Al_2O_3> Pt/Al_2O_3) for low methane conversions, and (3) similar asymptotic methane conversion levels of ~90% for temperatures above 550°C. The observation that the CH_4 conversion characteristics during a temperature run-up are essentially unaffected by CO in the feed is not surprising in view of the large disparity in their reactivity toward O_2. That is, much higher temperatures are required for methane oxidation than for CO oxidation, so that all the CO has been completely reacted by the time the CH_4 conversion becomes significant (see Figures 1 and 2).

For all three noble metal catalysts, methane oxidation under reducing conditions produced CO and H_2 as the principal products of partial oxidation. Our observation of H_2 as a reaction product during methane oxidation is perhaps not surprising in view of the report of Frennet ([20]) that at elevated temperatures methane chemisorption on noble metals is accompanied by H_2 evolution. Experiments conducted with feedstreams containing no CO were particularly convenient in examining the product distributions and the tendencies of the various noble metal catalysts to form CO during methane oxidation. Table 2 shows the amount of CH_4 consumed and the

Table 2. Product Distributions for Methane
Oxidation Under Reducing Conditions[a]

Catalysts	CH_4 Consumed (ppm)	CO_2 Produced (ppm)	CO Produced (ppm)	H_2 Produced (ppm)
Pt/Al_2O_3	1005	820	163	600
Pd/Al_2O_3	1000	800	250	715
Rh/Al_2O_3	1560	1025	560	2700

SOURCE: Reprinted with permission from ref. 35. Copyright 1991 Academic.

[a]0.2 vol% CH_4 and 0.15 vol% O_2 in the feed,
catalyst temperature = ~520°C, space velocity = 52 000 h^{-1}

amounts of CO, CO_2, and H_2 produced during methane oxidation under reducing conditions (0.2 vol% CH_4, 0.15 vol% O_2, and no CO in the feed) over the three noble metal catalysts at a catalyst temperature of ~520°C. For each of the noble metal catalysts, there is a good correlation between the amount of CH_4 reacted and the sum of the amounts of CO and CO_2 produced during the methane oxidation (carbon balance closure to within 5%), indicating that CO and CO_2 are the only carbon-containing reaction products under our experimental conditions. (The presence of elemental carbon on the surface is ruled out because O_2 exposure of the catalyst following steady-state reaction at low O_2/CH_4 ratios did not produce significant amounts of CO or CO_2.) For all the runs conducted under reducing conditions, including those presented in Table 2, the mass balance for oxygen agreed within 10% when CO, CO_2, H_2, and H_2O were assumed to be the only reaction products. Considering the uncertainty associated with the H_2 analysis (typical precision = ±5% of the reported value; our GC column was not optimized for H_2 analysis), it appears that no partial oxidation products other than CO and H_2 are formed in significant quantities under our reaction conditions. Although trace amounts of methanol, formaldehyde, and formic acid have also been reported as the reaction products in some cases (21), we did not pursue this point in detail.

The product selectivity to CO (i.e., the amount of CO produced divided by the amount of CH_4 consumed) for each noble metal can be calculated from the data of Table 2: 0.16 for Pt/Al_2O_3, 0.25 for Pd/Al_2O_3, and 0.36 for Rh/Al_2O_3. These calculated selectivities indicate that the tendency to form the partial oxidation product CO decreases in the order Rh/Al_2O_3 > Pd/Al_2O_3 > Pt/Al_2O_3, in agreement with the ranking of the noble metal catalysts for CO conversion observed under the reducing feedstream conditions of Figure 3. It is interesting to note that the H_2/CO ratio differs significantly among the noble metals: ~4 for Pt/Al_2O_3, ~3 for Pd/Al_2O_3, and ~5 for Rh/Al_2O_3. These compare with a H_2/CO ratio of 2 expected for the dissociative adsorption of methane leading to the cleavage of the C-H bond, a key step in methane oxidation over noble metals (7,10,11). This implies that for all three noble metals, the CO generated during the methane oxidation reacts more readily with oxygen than the H_2. In fact, the high H_2/CO ratio observed for the Rh/Al_2O_3 catalyst is consistent with the results of separate reactor experiments with CO-H_2-O_2 mixtures (22), which show that Rh/Al_2O_3 is particularly effective in preferentially oxidizing CO in the presence of a large excess of H_2.

Mechanistic Discussion

Under the oxidizing feedstream conditions considered here, the noble metal surface is predominantly covered with oxygen and thus the critical step in the methane oxidation is the dissociative adsorption (and subsequent reaction) of CH_4 onto the oxygen-covered surface (10,11,23-25). The supporting evidence for this conclusion is given below. First, most kinetic studies conducted over noble metal catalysts in the presence of excess O_2 show that the observed rate law is first order in methane and zero order in oxygen (11,19,24). Second, oxygen adsorption on noble metals is fast under the reaction conditions considered here (26), whereas methane adsorption is a slow, activated process (20). As a result of the substantial difference in adsorption strengths (O_2 > CH_4), O_2 inhibits the CH_4 oxidation under oxidizing conditions by excluding the more weakly adsorbed species, CH_4, from the active sites, as observed in Figure 3B. (The presence of CO in the feed would probably not significantly change the relative surface concentration of CH_4 and O_2 under reaction conditions, because CO tends to desorb rapidly from the surface at the high temperatures required for the methane reaction.)

As the O_2 concentration in the feed is decreased to less than the stoichiometric value (while holding the concentration of the reducing agents constant), however, CH_4 can compete successfully with O_2 for the active sites despite its weaker adsorption strength. In this case, the adsorption rates for both reactants would become comparable, yielding a high reaction rate and thus high CH_4 conversion. With a further decrease in inlet O_2 concentration, insufficient amounts of O_2 adsorb on the surface and the CH_4 conversion becomes low because of the resulting kinetic and/or stoichiometric limitations encountered during the methane oxidation. This explains why the CH_4 conversion goes through a maximum as the inlet O_2 concentration is varied, as illustrated in Figure 3B. A similar dependence of methane conversion on exhaust stoichiometry, including a maximum CH_4 conversion under somewhat richer than stoichiometric conditions, was observed over commercial noble metal catalysts in a recent engine-dynamometer study with methane fuel (27).

The detailed mechanism of methane oxidation on noble metals is not yet well understood. Methane chemisorption and methane-deuterium exchange experiments (20,28) have shown that the chemisorption of methane on noble metals involves dissociation to adsorbed methyl or methylene radicals, as a result of removal of hydrogen atoms from the carbon atom. The subsequent interaction of methyl or methylene radicals with adsorbed oxygen has been proposed to lead to either direct oxidation to CO_2 and H_2O or the formation of chemisorbed formaldehyde via methoxide, methyl peroxide, or methylene oxide intermediates (7,23,29,30). Recent studies of formaldehyde oxidation on Pt by McCabe and McCready (31), and by Lapinski et al. (32) provided kinetic and spectroscopic evidence that the oxidation reaction involves the dissociation of the adsorbed formaldehyde to adsorbed CO and adsorbed H atoms. These adsorbed CO and H atoms have been observed to either desorb as CO and H_2 or react with adsorbed oxygen to produce CO_2 and H_2O, depending on the stoichiometry of the reactant gas mixture (31,32). Based on the above discussion, the proposed reaction mechanism for methane oxidation can be represented by the following parallel-consecutive process.

The mechanism proposed above is consistent with our observation of CO, CO_2, H_2, and H_2O as the principal reaction products of the methane oxidation under reducing conditions. Although not directly measured, material balance considerations indicate that no significant amount of gaseous formaldehyde was produced during the methane oxidation over the noble metal catalysts. This suggests that under our reaction conditions an adsorbed formaldehyde intermediate, once formed, may rapidly decompose to CO(a) and H(a) rather than desorb into the gas phase as formaldehyde molecules.

Given the presence of the reaction products, CO, CO_2, H_2, and H_2O, in the reaction mixture, it is possible that the product distribution for the methane oxidation may be affected by the water-gas shift equilibrium reaction

$$CO_2 + H_2 \rightleftarrows CO + H_2O$$

(The equilibrium constant for the reaction as written above is 0.24 at 520°C.) A similar reaction mechanism has recently been proposed by Ashcroft et al. (33) for selective oxidation of methane to synthesis gas over transition metal catalysts. It should be noted that the product distributions given in Table 2 for actual reaction conditions deviate significantly from that dictated by the water-gas shift equilibrium, presumably because some of the CO and H_2 would react rapidly with O_2 in the stream. Nevertheless, one feature of the water-gas shift reaction is amenable to testing its importance in our system; that is, if the shift reaction is operative, the addition of CO_2 to the feed should increase CO production. To test this feature, we first established a steady-state reaction condition at 520°C with a reducing feedstream containing 0.2 vol% CH_4 and 0.15 vol% O_2 and then added CO_2 stepwise to the feedstream while holding the temperature and total flow rate constant. The results of such experiments for the three alumina-supported noble metal catalysts are presented in Table 3, which lists methane conversions and the amounts

Table 3. Methane Conversion and CO Production
With and Without CO_2 Addition

	Pt/Al$_2$O$_3$		Pd/Al$_2$O$_3$		Rh/Al$_2$O$_3$	
	CH_4 Conv. (%)	CO (ppm)	CH_4 Conv. (%)	CO (ppm)	CH_4 Conv. (%)	CO (ppm)
Base Case[a]	50	163	50	250	78	560
1000 ppm CO_2 Added	51	177	52	275	76	687
2000 ppm CO_2 Added	51	181	50	250	76	750

SOURCE: Reprinted with permission from ref. 35. Copyright 1991 Academic.

[a]0.2 vol% CH_4 and 0.15 vol% O_2 in the feed (no CO_2 added), catalyst temperature = ~520°C, space velocity = 52 000 h^{-1}

of CO produced during methane oxidation with and without CO_2 addition to the feed. (A more detailed description of the product distribution for the base case is given in Table 2.) It can be seen that for each of the noble metal catalysts the methane conversion is independent of the CO_2 level in the reaction mixture. The amount of

CO produced over either Pt/Al_2O_3 or Pd/Al_2O_3 is also virtually unaffected by the CO_2 addition. For the Rh/Al_2O_3 catalyst, on the other hand, increasing the CO_2 level in the reaction mixture tends to enhance CO production (and, though not shown, suppress H_2 production) during the methane oxidation. This suggests that the water-gas shift reaction plays a significant role in determining the product selectivity for methane oxidation over Rh/Al_2O_3, but not over Pt/Al_2O_3 or Pd/Al_2O_3. The insensitivity of the methane conversions over the noble metals to the CO_2 level in the feed suggests that the rate-limiting step in the methane oxidation, which is generally believed to be the dissociation of adsorbed methane molecules (10,11), is unaffected by CO_2.

An obvious question is the degree to which this water-gas shift argument might carry over into the actual exhaust environment. Vehicle exhaust contains a large amount of H_2O, which would strongly favor the reverse direction of the shift reaction as written above. However, CO_2 is also present in exhaust in large quantities, and thus the formation of CO from CO_2 cannot be ruled out. In addition, the presence of SO_2 (20 ppm by volume) in the feedstream has been shown to be highly detrimental to the water-gas shift reaction (34). However, the effect of SO_2 on the performance of natural-gas exhaust catalysts is expected to be small since analysis of typical compressed natural gas shows that its SO_2 content is less than 5 ppm by volume.

Literature Cited

1. Weaver, C. S. SAE Paper 1989, No. 892133.
2. Martin, W. F.; Campbell, S. L. "Natural Gas: A Strategic Resource for the Future," Washington Policy and Analysis, Washington, D.C., 1988.
3. DeLuchi, M. A.; Johnston, R. A.; Sperling, D. SAE Paper 1988, No. 881656.
4. Golomb, D.; Fay, J. A. "The Role of Methane in Tropospheric Chemistry," Energy Laboratory, Cambridge, MA, 1989.
5. Hillemann, B. Chemical and Engineering News 1989, 67 (11), 25.
6. Cadle, S. H.; Mulawa, P. A.; Hilden, D. L.; Halsall, R. "Exhaust Emissions from Dual-Fuel Vehicles Using Compressed Natural Gas and Gasoline", presented to the Air and Water Management Association, Pittsburgh, PA, June 1990.
7. Golodets, G. I. "Heterogeneous Catalytic Reactions Involving Molecular Oxygen," in Stud. Surf. Sci. and Catal. Vol. 15; Elsevier: Amsterdam, 1983; Chapter XV.
8. Anderson, R. B.; Stein, K. C.; Feenan, J. J.;Hofer, L. J. E. Ind. Eng. Chem. 1961, 53, 809.
9. Firth, J. G.; Holland, H. B. Faraday Society Transactions 1969, 65, 1121.
10. Yu Yao, Y.-F. Ind. Eng. Chem. Prod. Res. Dev. 1980, 19, 293.
11. Otto, K. Langmuir 1989, 5, 1364.
12. Schlatter, J. C.; Taylor, K. C.; Sinkevitch, R. M. "The Behavior of Supported Rhodium in Catalyzing CO and NO Reactions," presented at Adv. in Catal. Chem. Symp., Snowbird, UT, October 1979.
13. Dictor, R.; Roberts, S. J. Phys. Chem. 1989, 93, 2526.
14. Oh, S. H.; Eickel, C. C. J. Catal. 1988, 112, 543.

15. McCabe, R. W.; Mitchell, P. J. Ind. Eng. Chem. Prod. Res. Dev. 1984, 23, 196.
16. Voltz, S. E.; Morgan, C. R.; Liederman, D.; Jacob, S. M. Ind. Eng. Chem. Prod. Res. Dev. 1973, 12, 294.
17. McCabe, R. W.; Mitchell, P. J. Appl. Catal. 1986, 27, 83.
18. Kummer, J. T. Prog. Energy Combust. Sci. 1980, 6, 177.
19. Niwa, M.; Awano, K.; Murakami, Y. Appl. Catal. 1983, 7, 317.
20. Frennet, A. Catal. Rev.-Sci. Eng. 1974, 10, 37.
21. Margolis, L. Y. Adv. Catal. 1963, 14, 429.
22. Sinkevitch, R. M.; Oh, S. H. "A Method and Apparatus for Selective Removal of Carbon Monoxide," U.S. Patent filed, SN#709,563, 1991.
23. Cullis, C. F.; Keene, D. E.; Trimm, D. L. J. Catal. 1970, 19, 378.
24. Cullis, C. F.; Willatt, B. M. J. Catal. 1983, 83, 267.
25. Hicks, R. F.; Qi, H.; Young, M. L.; Lee, R. G. J. Catal. 1990, 122, 280.
26. Engel, T.; Ertl, G. Adv. Catal. 1979, 28, 1.
27. Siewert, R. M.; Mitchell, P. J., General Motors Research Laboratories, Warren, MI, unpublished data, 1990.
28. Kemball, C. Adv. Catal. 1959, 11, 223.
29. Dowden, D. A.; Schnell, C. R.; Walker, G. T. Proc. 4th Int. Congr. Catal., Moscow, 1968, p. 1120.
30. Pitchai, R.; Klier, K. Catal. Rev.-Sci. Eng. 1986, 28, 13.
31. McCabe, R. W.; McCready, D. F. Chem. Phys. Letts. 1984, 111, 89.
32. Lapinski, M. P.; Silver, R. G.; Ekerdt, J. G.; McCabe, R. W. J. Catal. 1987, 105, 258.
33. Ashcroft, A. T.; Cheetham, A. K.; Foord, J. S.; Green, M. L. H.; Grey, C. P.; Murrell, A. J.; Vernon, P. D. F. Nature 1990, 344, 319.
34. Schlatter, J. C.; Mitchell, P. J. Ind. Eng. Chem. Prod. Res. Dev. 1980, 19, 288.
35. Oh, S. H.; Mitchell, P. J.; Siewert, R. M. J. Catal. 1991, 132, 287.

RECEIVED January 3, 1992

Chapter 3

Automotive Catalyst Strategies for Future Emission Systems

W. Burton Williamson, Jerry C. Summers, and John A. Scaparo

Allied-Signal Automotive Catalyst Company, P.O. Box 580970, Tulsa, OK 74158—0970

While significant advances in Pt/Rh three-way catalyst (TWC) formulations have been accomplished, the use of Pd-containing catalysts for three-way emission control are of interest for overall noble metal cost reduction, lower Rh usage, and potential durability improvements. Applications of Pd are demonstrated for replacement of Pt in conventional Pt/Rh TWC systems, for use in Pd-only three-way catalysts and for lowering methanol and formaldehyde emissions at close-coupled locations on a methanol-fueled vehicle. The individual contributions of Pt, Pd and Rh for aged three-way performance indicate significant advantages of using Pd over Pt. A comparison of vehicle system control strategies illustrates that higher   system temperatures significantly lower HC emissions, while air/fuel control strategies are most critical in lowering NOx emissions.

Advanced automotive catalyst technologies and vehicle emission systems will be critical in achieving the more stringent NOx and hydrocarbon emission standards of the 1990s that are being implemented world-wide. Future automotive emission systems will require further significant advancements in order to help improve national as well a global air quality, since the global vehicle population has increased ten-fold in the last forty years to over 500 million vehicles(1,2). Projections based on linear regressions at an average rate of 2.5% per year would double worldwide vehicle growth to one billion vehicles over the next forty years(3).

Since passenger car emission standards established by the Clear Air Act of 1970 and implemented beginning model year 1975 in the United States generally required catalysts, there has been increasing pressure to decrease emissions further. Compared to uncontrolled emission levels prior to the enactment of the Clean Air Act, current federal exhaust emission levels (Table 1) have reduced hydrocarbon (HC) and carbon monoxide emissions by 95% and

0097—6156/92/0495—0026$06.00/0

NOx emissions by 70%(4). Further requirements increase the potential warranty from 50,000 miles to 100,000 miles (80,000 to 160,000km) in addition to the lower emission levels discussed in Chapter 1.

California adopted a stringent NOx standard of 0.4g/mi beginning with the 1989 model year. The feasibility of simultaneously achieving 0.4g/mi NOx and 0.25g/mi HC standards is based upon certification results that indicate an estimated 40% of 1987 California certified cars are capable of meeting the proposed future emission standards. The average 1986 and 1987 California cars and light trucks had certification levels of 0.20g HC, 2.9g CO and 0.4g NOx/mile(5).

Catalytic NOx control is primarily dependent on the amount of rhodium (Rh) noble metal used and the type of catalyst system. In California the relatively low levels of NOx was largely achieved by using only TWCs, rather than dual-bed systems. Typically the TWC-only systems used about twice as much Rh as the dual-bed systems, and to achieve the 0.4 NOx standard required even significantly larger Rh loadings than for the 0.7 NOx standard. However, factors such as expensive Rh, scarcity, low recycling recovery rate, and increased global usage are not conducive to increased vehicle per capita consumption.

On the contrary, during the past few years noble metal cost reduction pressures have generated significant interest in the use of Pd-containing catalysts for three-way emission control applications(6-11). In addition, Pd/Rh catalysts have the potential of improved hydrocarbon emission performance(8).

Major improvements of catalytic control of HC emissions can be made by significantly reducing the time required for the catalyst to light-off and reach operating temperature, since up to 75% of the engine-out HC emissions for a typical vehicle can occur during the cold-start portion (Bag 1) of the FTP-75 cycle. One way to achieve this is by the use of close-coupled three-way catalysts. However, high exhaust temperatures result in thermal deterioration of closely mounted catalysts which must be minimized to maintain extended durability, especially for 100,000 mile considerations.

The durability of current automobile catalysts in the U.S. is governed primarily by thermal deactivation(12), since chemical deactivation has been greatly reduced by the use of unleaded gasoline. The extent of performance loss is not only a function of the maximum temperature of catalyst exposure, but also the air/fuel (A/F) range of the exhaust during high temperature exposure.

Various catalyst strategies will need to be considered in order to meet the proposed emission standards. One approach is to increase the noble metal content of the catalyst as well as perhaps increasing the catalyst volume. A second is to use three-way catalysts closely coupled to the engine manifold in order to insure superior light-off performance. A third is to use start-up catalysts near the manifold coupled with an underfloor catalyst (including by-pass catalysts). Another option would be to use conventional underfloor catalyst technology having improved formulations.

In order to illustrate the strengths and weaknesses of each of these approaches, selected experimental data will be presented that was generated in the course of development of newer catalyst technologies. Improvements only in catalyst technologies may not be sufficient to enable the automotive engineer

to meet increasing stringent emission targets.   However, new catalyst
technologies currently under development coupled with adequate engine fuel
management systems and optimized calibrations will allow the car companies
to   make   substantial   progress   toward   meeting   the   future
TLEV/LEV/ULEV/ZEV emission standards discussed in chapter 1.

Improving Catalyst Performance

Increased Noble Metal Content. The most obvious (and most expensive)
approach for meeting more stringent HC and NOx standards is to increase the
noble metal content of the catalyst. Such approaches have proven successful as
indicated by California certification data[5].   However, higher three-way
conversions and lower light-off temperatures have been demonstrated at
constant noble metal content by placing Rh nearer the front of the catalyst bed
rather than uniformly throughout the bed[13].

The sensitivity of light-off performance to increasing noble metal loading
at a constant Pt/Rh weight ratio (5/1) is shown in Figure 1[14]. Catalysts in
these experiments were aged for 100 hours on a fuel-cut aging cycle (maximum
catalyst inlet temperature = 760°C) with fuel containing 15mg lead/gal and 2mg
phosphorus/gal (4mg Pb and 0.5mg P/liter).

Cerium Promoters/Stabilizers. Base metal promoters, such as cerium
(Ce), are incorporated into three-way catalyst formulations to significantly
enhance NOx and CO performance near stoichiometric A/F ratios[15,16]. The
beneficial effects of Ce on improving the stoichiometric performance of Pt/Rh
TWCs (10/1 at 0.24 wt% = 40g/ft^3) are shown in Figure 2[14] after thermal
aging at 900°C for 4h with 10% H_2O in air. Catalyst loadings are reported in
wt%, or g/ft^3 of ceramic monolith substrate having 64 square cells/cm^2 (400
cells/in^2).   Increasing Ce loadings (at constant total washcoat loadings)
significantly improves the NOx and CO conversions at 550°C with only a slight
effect on HC oxidation.

The improved NOx conversions result from less thermal deactivation of
Rh, which occurs during high-temperature lean aging and has been attributed
to strong Rh-alumina interactions[17,18]. During cyclic operation around the
exhaust gas stoichiometric point, additional CO can be removed by the water-
gas shift reaction over Pt[19,20] and Rh[19]. An enhancement of the water-gas
shift reaction occurs with increasing Ce loadings[21].   Light-off durability
performance also improves with increasing cerium loadings[7].

Individual Noble Metal Durability. The individual contributions of Pt,
Pd and Rh for aged three-way performance were compared to Pt/Rh and
Pd/Rh TWCs using a high performance commercial washcoat technology
previously optimized for Ce-containing Pt/Rh catalysts. The Pt/Rh and Pd/Rh
TWC contained nominal loadings of 0.2 wt% Pt or Pd (33.3g/ft^3) and 0.04 wt%
Rh (6.7g/ft^3), and the single component catalysts were formulated at the same
respective loadings, i.e., 0.2 wt% Pt, 0.2 wt% Pd and 0.04 wt% Rh. In addition,
all catalysts contained equivalent proprietary base-metal and washcoat loadings.

Catalysts were aged simultaneously in dual quadrant reactors for 100 hr
on a severe fuel-cut aging cycle (maximum catalyst inlet temperature = 850°C)

TABLE 1

PASSENGER CAR EMISSION STANDARDS

MODEL YEAR	Federal Test Procedures (g/ml)		
	HC	CO	NOx
Uncontrolled	10.6	84	4.1
1970	4.1	34	NR
1972	3	28	NR
1973	3	28	3.1
1975	1.5	15	3.1
1977	1.5	15	2.0
1980	0.41	7	2.0
1981	0.41	3.4	1.0
1993 California (50k)	0.25*	3.4*	0.4
(100k)	0.31	4.2	---

*% Compliance: 40% 1993, 80% 1994, 100% 1995

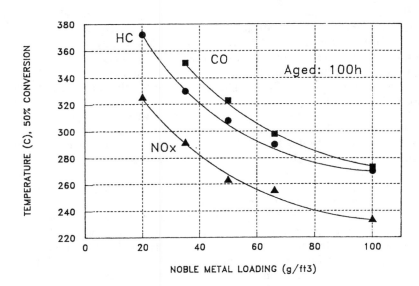

Figure 1. Effect of noble metal loadings on light-off temperatures of Pt/Rh (5/1) catalysts after engine aging. (Reproduced with permission from ref. 14. Copyright 1988 Society of Automotive Engineers, Inc.)

using commercial unleaded fuel. Results of the individual evaluations at 450°C are shown in Figure 3 as integral stoichiometric conversions (A/F = 14.56 ± 0.15).

At stoichiometric A/F ratios, the Pd/Rh and Pt/Rh catalysts are equivalent, while the Pt/Rh TWC had slight 5% NOx conversion advantages at both 2% lean and 2% rich A/F conditions. The relative performance rankings in the range of A/F = 14.25 (2% rich) to 14.85 (2% lean) indicate the following:

Pt/Rh = Pd/Rh > Pd > Rh >> Pt for HC, CO
Pt/Rh ≥ Pd/Rh > Rh > Pd >> Pt for NO

The only significant contribution for Pt was for lean HC/CO conversions, which were still lower than the Pd-Ce catalyst. These results indicate significant advantages of using Pd-Ce over Pt-Ce catalysts for oxidation reactions and NO reduction, however, the predominant contribution for the three-way performance comes from Rh, even though Rh is at much lower loading. Indeed substantial removal of Pt from Pt/Rh TWCs can be achieved before significantly affecting stoichiometric conversions or light-off performance of Pt/Rh TWCs.(13).

Increased Palladium Usage

Pd/Rh TWC Performance. Recent studies illustrate the progress that has been made in developing high performance Pd/Rh catalyst technologies(11). In previous studies evaluating commercial Pt/Rh and Pd/Rh technologies, the Pd/Rh TWC had lower NOx conversions and required relatively tight A/F control to approach the performance levels achieved by the Pt/Rh TWC(10).

Despite price and availability advantages, historically there have been a number of problems connected with Pd usage. These included alloy formation during use between Rh and Pd(22,23), palladium's ability to alloy with Pt(24), lower activity in oxidizing C_3 or higher molecular weight paraffins(25), and its susceptibility towards poisoning from fuel and oil contaminants(26). These problems have now largely been overcome and a variety of high performance Pd/Rh catalyst technologies are currently being introduced into the marketplace. On the positive side, Pd has well-known durability performance advantages relative to Pt: in oxidizing environments, it is considerably more resistant to sintering than Pt(24,27).

For Pd/Rh and Pt/Rh catalysts at identical loadings that were aged under five different high temperature aging cycles, Figure 4(11) illustrates the differences in conversion at the stoichiometric point between the aged catalyst pairs. Positive differences indicate higher conversions for the Pd/Rh catalyst, while negative differences indicate higher conversions for the Pt/Rh catalyst. The conversion performance data reveal far more similarities than differences for the two noble metal technologies. Conversion differences of less than 4% are not considered statistically significant.

Pd-Only Three-Way Control Catalysts. Rh is the component of current conventional three-way control catalysts responsible for converting NOx to N_2

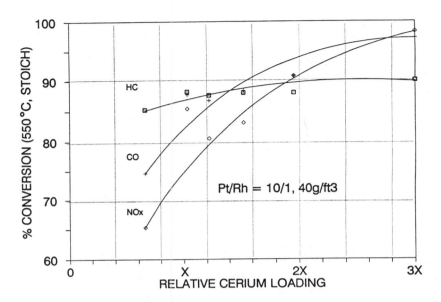

Figure 2. Effect of Ce loading on Pt/Rh catalyst conversions after 900 °C aging with 10% H_2O in air. (Reproduced with permission from ref. 14. Copyright 1988 Society of Automotive Engineers, Inc.)

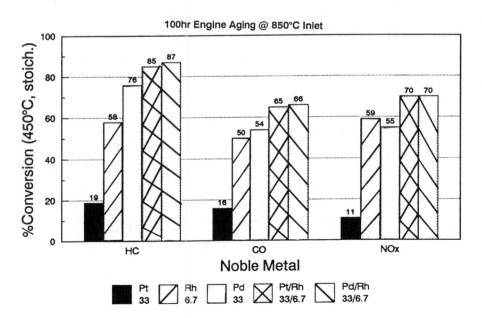

Figure 3. Noble metal catalyst durability after engine aging at 850 °C maximum inlet temperature.

by its reaction with CO(28). Pd also has some ability to convert NOx(29,30). The use of Pd as a three-way control catalyst in a single bed application, as a replacement for Pt/Rh in the rear bed of a two-bed Pt/Rh three-way control system, and as a catalyst incorporated into a two bed Pd-only dual bed system with secondary air injected between the two catalyst beds has been examined(7,8).

Figure 5(7) gives the durability performance at the stoichiometric point of Pd-only three-way catalysts as a function of Pd loading after aging for 100h on a fuel-cut aging cycle. The HC performance is relatively independent of Pd loading (0.12-0.60 wt%/20-100g/ft^3) ranging from 88% to 93%, respectively. The HC conversion levels are high and compare well with those achieved by high-tech Pt/Rh catalysts aged under similar conditions. This may appear surprising since Pd is thought to be inferior to Pt in its ability to convert difficult-to-oxidize hydrocarbons such as propane(25) which may comprise 15-20% of the hydrocarbons normally in automotive exhaust(31). However, as these results clearly show, Pd can be used with a proper washcoat technology to give high levels of HC control near the stoichiometric point. The NOx conversions are also relatively high and strongly dependent on Pd loading, ranging from 49% to 69% (0.12 to 0.60 wt%). The CO conversions exhibit similar dependency on Pd content.

Improved durability performance was also noted for all regulated emissions with increasing Pd content except for lean NOx and rich CO conversions which were essentially constant in A/F traverse evaluations. Also, the temperature required to convert 50% of the HC, CO and NOx over the Pd is almost directly related to Pd-loading: Pd catalysts with 0.6 wt% ≈ 100g/ft^3 loadings have about the same performance as 5/1 Pt/Rh catalysts at 0.24 wt% (40g/ft^3) when aged under similar conditions.

Typically, lead poisoning has a greater impact on HC conversion than CO conversion for noble metal catalysts under lean (27,32) and rich (26,33) operation. More recent Pd formulations have been found to maintain a performance advantage with respect to Pt/Rh catalysts for HC conversion regardless of lead fuel level(8). The effect of lead poisoning for both catalysts is particularly evident rich of the stoichiometric point. Lower rich A/F CO performance was found for Pd catalysts than for Pt/Rh catalysts. Near the stoichiometric point (≈ 14.45-14.70) there is little impact of lead poisoning for the HC or CO conversion of either catalyst.

One general feature of the NOx conversion for the Pd TWC is that it rapidly falls off rich of the stoichiometric point relative to the Pt/Rh catalyst. Part of this difference is due to the higher NH$_3$ formation found over Pt/Rh catalysts(7). Most closed-loop vehicles operate very near the stoichiometric point where the Pd TWC performs well. If improvements in rich NOx conversion are desired, small quantities of Rh can be added, as shown in Figure 6(8).

For NOx conversion, there is also a greater effect of lead poisoning rich of the stoichiometric point than at the stoichiometric point. At the stoichiometric point, the Pd catalyst is superior to the Pt/Rh catalyst after exposure to lead at either level. Rich of the stoichiometric point, the NOx

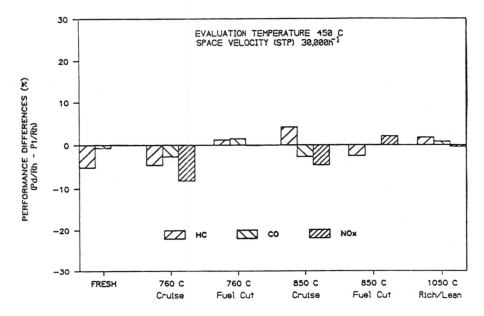

Figure 4. Pd/Rh and Pt/Rh catalyst comparisons for stoichiometric A/F at 450
° C after various accelerated engine aging cycles. (Reproduced with permission
from ref. 11. Copyright 1990 Society of Automotive Engineers, Inc.)

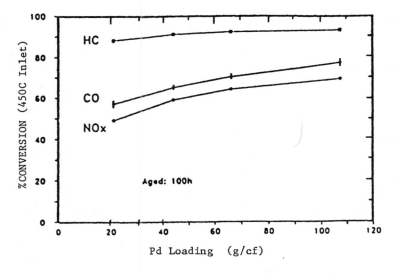

Figure 5. Effect of Pd loading on stoichiometric conversions at 450 ° C of Pd-
only catalysts after engine aging. (Reproduced with permission from ref. 7.
Copyright 1988 Society of Automotive Engineers, Inc.)

conversion of both catalysts falls off sharply upon exposure to increasing lead levels. In the poisoning of Pt/Rh catalysts, the NOx conversion was significantly deteriorated under strongly rich conditions, particularly the conversion of NOx to NH_3(27,33). Similar results were found over Pd catalysts(26).

Evaluation of Vehicle-Aged Pd Catalysts. Pd and Pt/Rh catalysts were aged for 25,000 miles (40,000km)on 1988 Dodge Dynastys, removed from the vehicles and evaluated on an engine dynamometer and across different vehicles according to the U.S. FTP-75 test. The FTP-75 results for the 25,000 mile aged Pd catalyst in Table 2(8) indicate that this technology has comparable durability performance to conventional high-tech Pt/Rh technology. Because vehicle aging data can be highly variable, some care should be exercised in the direct comparison of one set of vehicle performance numbers with another for the two different catalysts. However, the 25,000 mile data indicate that Pd-only TWC catalysts are viable.

The performance of the Pd catalyst is excellent across a relatively clean 2.5L Somerset (engine-out emissions are 1.2g HC/mi, 6.8g CO/mi, 1.5g NOx/mi). FTP-75 tailpipe emissions are very low: 0.07g HC/mi, 0.68g CO/mi and 0.36g NOx/mi. These findings support the idea that Pd-only TWC technology is excellent for well-designed tight emission control/fuel management systems. Furthermore, the FTP-75 conversions of the Pd catalyst aged on a driving cycle more severe than the AMA driving cycle exceed those established by the EPA (70% HC/ 70% CO/30%NOx) and the Air Resources Board of California (70% HC/70% CO/50% NOx) for meeting aftermarket catalytic converter performance standards. Considering the conversions of the Pd catalyst measured across the relatively cold Nissan 300ZX (80% HC/ 81% CO/64%NOx), it is obvious that this technology can easily meet the aftermarket requirements for vehicles of 3500 lbs (1588kg) GVW or lighter. In further aging to 50K miles (80K km) the Pd TWC deteriorated further, but still easily met U.S. standards.

Palladium Usage in Methanol Vehicles. Methanol fueled vehicles will be required to meet a 15mg/mile formaldehyde (HCHO) emission standard in California starting in 1993, i.e. levels comparable to formaldehyde emissions from current three-way catalyst equipped vehicles. Reportedly Pt and Pd catalysts are good for high methanol conversions with low formaldehyde emissions, and lowest formaldehyde emissions of Pt and Pd based TWCs occur during stoichiometric or slightly rich conditions(34).

Methanol and formaldehyde emissions over Pd and Pt/Rh close-coupled catalysts were determined as part of a California Air Resources Board emission control program at Southwest Research Institute using a 1981 gasoline Ford Escort modified to operate on 90% methanol fuel (M90). A developmental Pd catalyst at 0.29 wt% ($50g/ft^3$) and a high performance commercial Pt/Rh = 5/1 at 0.12 wt% ($20g/ft^3$) TWC (both at 1.0 liter volumes) were evaluated fresh in a close-coupled location (76cm from manifold) and compared to the stock converter of the 1981 gasoline Escort (TWC + oxidation) located in an underbody location (183cm from manifold).

The 0.29 wt% catalyst had equivalent or better HC performance compared to the Pt/Rh TWC, and both close-coupled formulations had better

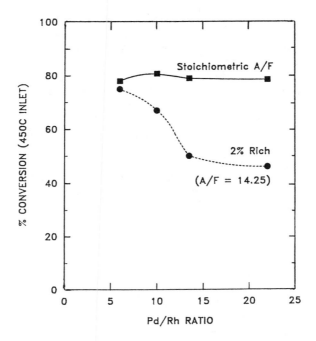

Figure 6. Effect of Rh content in Pd/Rh catalysts on NOx conversions at 450 °C after engine aging (760 °C inlet fuel-cut cycle). (Reproduced with permission from ref. 8. Copyright 1989 Society of Automotive Engineers, Inc.)

TABLE 2

VEHICLE EVALUATION OF 25K MILE AGED CATALYSTS
FTP-75

	Conversions (%)			
Modal Summary	**2.5L TBI Somerset**		**3.0L MPFI 300ZX**	
Total FTP - 75	**Pd**	**Pt/Rh**	**Pd**	**Pt/Rh**
HC	94	93	80	78
CO	90	92	81 > 77	
NOx	76 > 69		64 < 68	

SOURCE: Reprinted with permission from ref. 8. Copyright 1989 Society of Automotive Engineers, Inc.

HC conversions than the underbody Escort converter, as shown in Figure 7 for cold start, hot start, 30 mph and 50 mph (48 and 80km/h) testing. Also, both of the close-coupled TWCs had better three-way control in general than the underbody Escort converter.

Both Pd and Pt/Rh technologies located in the close-coupled position also had significantly lower methanol and formaldehyde emissions than the Escort TWC + COC underbody converter, as shown in Table 3. In cold start testing, methanol and formaldehyde emissions of the close-coupled systems were approximately 10% of the Escort system due to the significantly higher conversions. Close-coupled catalyst emissions during hot start were still only about 20% of the Escort emissions. Similar results are shown for the 30 and 50 mph testing. At the high conversion levels, little distinction could be made of a preferred technology since they were both comparable for methanol/formaldehyde emissions and durability considerations were not addressed.

Vehicle System Effects

Light-off Performance. Catalyst light-off performance is strongly related to A/F and improves dramatically over noble metal catalysts upon moving from rich to lean A/F ratios(27). Light-off is also strongly related to noble metal loading and noble metal composition(35).

A problem facing catalyst development scientists is how to measure light-off performance. As an aid to answering this question, it is instructive to consider how a vehicle operates during the cold transient phase of the FTP-75 cycle. Catalyst light-off typically occurs during the first 200 seconds of this test. The exact time that light-off occurs will vary and depends on catalyst composition factors and vehicle control strategies. The exhaust temperature ramp rate, A/F control strategy, and the time required to switch from open to closed-loop operation vary demonstratively for different vehicles. As a result, time required for HC light-off to occur is system dependent, as illustrated in Figure 8(11).

With the conventional three-way control systems (those with no secondary air added), there is a functional balance of temperature ramp rate and the time required to switch from open to closed-loop control. The closed-loop switch control point occurs only after the design operating temperature of the oxygen sensor has been reached ($>260°C$), the timing of which is highly dependent on the vehicle control system. Light-off occurs at some minimum inlet temperature after closed-loop operation is attained.

Vehicle Results. Improved HC performance levels correlate best with higher average reactor inlet temperatures. HC FTP-75 test results for an advanced Pt/Rh TWC are shown in Table 4(36) for 850°C fuel-cut aged samples across four different vehicles. The system differences of these vehicles with respect to fuel control, catalyst operating temperatures, and A/F environment of the catalyst have been described previously(36). Carbon monoxide performance results for the same catalyst formulation were consistently better for vehicles with less rich excursions. The lowest hydrocarbon performance

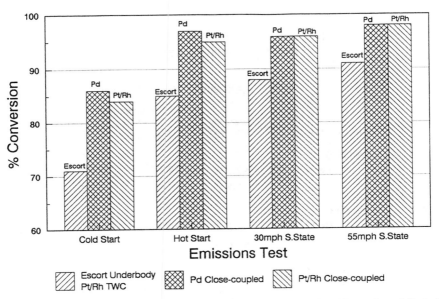

Figure 7. Hydrocarbon emission catalyst comparison of M90 methanol-fueled Escort vehicle.

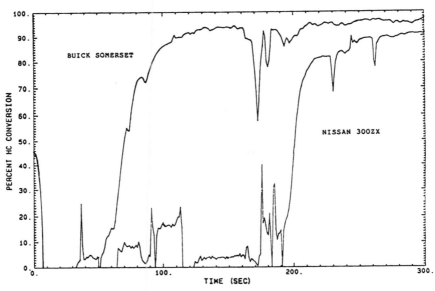

Figure 8. Effect of vehicle system on the time required for HC light-off to occur for a Pt/Rh catalyst after engine aging (850 °C inlet fuel-cut cycle). (Reproduced with permission from ref. 11. Copyright 1990 Society of Automotive Engineers, Inc.)

TABLE 3

METHANOL AND FORMALDEHYDE EMISSIONS
FROM M90 FUELED ESCORT

	EMISSIONS: mg/ml (% Conv.)		
	Escort Underbody	Developmental Close-Coupled	
Emission Test	Pt/Rh TWC	Pd	Pt/Rh
Cold Start			
MeOH	1300 (84)	170 (98)	150 (98)
HCHO	62.6 (80)	6.1 (98)	5.9 (98)
Hot Start			
MeOH	30 (99)	ND (99+)	ND (99+)
HCHO	9.4 (98)	2.1 (99)	1.4 (99)
30 mph Cruise			
MeOH	10 (99)	10 (99)	10 (99)
HCHO	9.5 (93)	2.1 (99)	1.3 (99)
55 mph Cruise			
MeOH	ND (99+)	ND (99+)	ND (99+)
HCHO	11.3 (96)	1.4 (99+)	1.2 (99+)

(ND = Not Detected)

TABLE 4

HC FTP PHASE SENSITIVITY TO VEHICLE SYSTEM
OF HIGH-TECH TWC[a]

	% HC Conversion			
	(Phase Avg. Catalyst Inlet T, °C)			
Vehicle	Total	CT (T°C)	CS (T°C)	HT (T°C)
Scirocco	84	72 (506)	91 (488)	81 (500)
Somerset	81	72 (435)	89 (410)	73 (405)
Citation	72	68 (408)	77 (372)	62 (392)
300ZX	63	36 (341)	81 (367)	61 (362)

a = TWC aged 100h 850°C Fuel-Cut Cycle

levels were obtained with Citation and 300ZX vehicles and highest with the Somerset and Scirocco vehicles. This substantiates methods to reduce HC emissions by increasing catalytic reactor temperatures, such as a manifold mount system.

Nitric oxide performance levels correlate best with the mean A/F control value during key high emission modes, i.e., accelerations and high speed cruises, as shown in Table 5(36). The net rich fuel calibration of the Scirocco vehicle, especially at high speeds, resulted in the highest NO conversions. The corresponding net lean calibration of the 300ZX vehicle resulted in the lowest NO conversions.

TABLE 5

NOx FTP ADVANCED TWC SENSITIVITY TO VEHICLE SYSTEM[a]

| Phase | Mode | % NOx Reduction From Phase Total[b] | | | |
		300ZX	Citation	Somerset	Scirocco
CT	Idle	1	2	4	1
	Accel.	13	18	20	22
	Cruise	18	41	39	60
	Decel.	2	3	3	6
	Total CT	34	64	66	89
CS	Idle	3	3	8	1
	Accel.	36	35	33	37
	Cruise	27	38	30	45
	Decel.	8	10	12	16
	Total CS	74	86	83	99
HT	Idle	2	1	3	1
	Accel.	25	15	18	26
	Cruise	25	46	51	62
	Decel.	3	3	5	6
	Total HT	55	65	77	95
Weighted Total Performance		-- 58	-- 74	-- 76	-- 95
A/F HT Hill 2 Cruise		14.72	14.52	14.56	14.38

a = Advanced TWC aged 100h on 850°C fuel-cut cycle

$$b = \frac{\text{Mode (Inlet Grams – Outlet Grams)}}{\text{Phase Total Grams}} \times 100\%$$

SOURCE: Reprinted with permission from ref. 36. Copyright 1989 Society of Automotive Engineers, Inc.

In an attempt to understand the origin of these significant differences, the catalyst warm-up and duration in which the fuel system switched to close-loop control was determined. For the Scirocco vehicle, the catalyst reached warm-up as early as the first hill. For the Nissan 300ZX, this was achieved very slowly and only during hill two of Bag 1. For the Citation vehicle, a moderately rapid warm-up was observed which occurred right after hill one. Concomitantly, the switch to closed-loop control was rapid on the Scirocco, very slow on the 300ZX and moderately rapid on the Citation.

Also important is the fact that the control set point during closed-loop operation for the Scirocco was at the stoichiometric or just slightly rich of it, whereas the 300ZX was set lean of the stoichiometric point. The Citation's control point was at the stoichiometric point. The relative air-to-fuel modulations were small for the multi-point fuel injection vehicles (Scirocco and 300ZX) and relatively large for the Citation vehicle. Consequently, the NOx conversion performance on the Scirocco was high for the warm cruise mode of Bag 1 and 3, low for the 300ZX and moderately high for the Citation.

Acknowledgements

L. A. Butler, P. M. Caouette, J. R. Coopmans, J. E. Dillon, K. J. Foley, M. G. Henk, D. G. Linden, P. J. Pettit, J. R. Provenzano, R. J. Shaw, J. F. Skowron, M. Thomason, D. Whisenhunt, J. H. White.

Literature Cited

1. Walsh, M.P.; Karlsson, J.; SAE (Society of Automotive Engineers) 1990, Paper 900613.
2. MVMA (Motor Vehicle Manufacturers Assoc.) Motor Vehicle Facts and Figures '90, 1990, p.36.
3. Walsh, M.P.; Platinum Metals Rev. 1989,33,194.
4. U.S. Environmental Protection Agency, "Mobile Source Emission Standards Summary", March 20, 1985.
5. Lyons, J.M.; Kenny, R.J.; SAE 1987, Paper 872164.
6. Muraki, H.; Shinjoh, H.; Sabukowa, H.; Yokata, K.; Fujitani, Y.; Ind. Eng. Chem. Prod. Res. Dev. 1986, 25, 202.
7. Summers, J.C.; Williamson, W.B.; Henk, M.G.; SAE 1988, Paper 880281.
8. Summers, J.C.; White, J.J.; Williamson, W.B.; SAE 1989, Paper 890794.
9. Gandhi, H.S.; Watkins, W.L.H.; U.S. Patent 4,782,038; November 1, 1988.
10. Heck, R.; Patel, K.S.; Adomaitis, J.; SAE 1989, Paper 892094.
11. Summers, J.C.; Williamson, W.B.; Scaparo, J.A.; SAE 1990, Paper 900495.
12. Otto, K.; Williamson, W.B.; Gandhi, H.S.; Ceramic Eng. and Sci. Proc. 1981, 2, 352.
13. Summers, J.C.; Hiera, J.P.; Williamson, W.B.; SAE 1991, Paper 911732.

14. Williamson, W.B.; Summers, J.C.; Skowron, J.F.; SAE 1988, Paper 880103.
15. Kim,G.; Ind. Eng. Chem. Prod. Res. Dev. 1982, 21, 267.
16. Cooper, B.J.; Truex, T.J.; SAE 1985, Paper 850128.
17. Yao, H.C.; Japar, S.; Shelef, M.; J. Catal. 1977, 50, 407.
18. Yao, H.C.; Stepien, H.K.; Gandhi, H.S.; J. Catal. 1980, 61, 547.
19. Joy, G.C.; Lester, G.R.; Molinaro, F.S.; SAE 1979, Paper 790943.
20. Gandhi, H.S.; Piken, A.G.; Stepien, H.K.; Shelef, M.; Delosh, R.G.; Heyde, M.E.; SAE 1977, Paper 770196.
21. Cooper, B.J.; Evans, W.D.J.; Harrison, B.; in Catal. Automot. Pollut. Control; Cruq, A., Ed.; Stud. Surf. Sci. Catal. Vol. 30; Elsevier: Amsterdam, 1987.
22. Graham, G.W.; Potter, T.; Baird, R.J.; Gandhi, H.S.; Shelef, M.; J. Vac. Sci. Technol. 1986, 4, 1613.
23. Hegedus, L.L.; Summers, J.C.; Schlatter, J.C.; Baron, K.; J. Catal. 1979, 54, 321.
24. Summers, J.C.; Hegedus, L.L.; J. Catal. 1978, 51, 185.
25. Yao, Y.F.; Ind. Eng. Chem. Prod. Res. Dev. 1980, 19, 293.
26. Williamson, W.B.; Lewis, D.; Perry, J.; Gandhi, H.S.; Ind. Eng. Chem. Prod. Res. Dev. 1984, 23, 531.
27. Summers, J.C.; Monroe, D.R.; Ind. Eng. Chem. Prod. Res. Dev. 1981, 20, 23.
28. Bauerle, G.L.; Service, G.R.; Nobe, K.; Ind. Eng. Chem. Prod. Res. Dev. 1972, 11, 54.
29. Schlatter, J.C.; Taylor, K.C.; J. Catal. 1977, 49, 42.
30. Gandhi, H.S.; Yao, H.C.; Stepien, H.K.; ACS Symposium Series 1982, 178, 143.
31. Jackson, M.W.; SAE 1978, Paper 780624.
32. Shelef, M.; Dalla Betta, R.A.; Larson, J.A.; Otto, K.; Yao, H.C.; "Poisoning of Monolithic Noble Metal Oxidation Catalysts in Automobile Exhaust Environment", 74th National Meeting of AIChE, New Orleans, March 1973.
33. Williamson, W.B.; Gandhi, H.S.; Heyde, M.E.; Zawacki, G.A.; SAE Transactions 1979, 88, 3196 (Paper 790942).
34. McCabe, R.W.; King, E.T.; Watkins, W.L.H.; Gandhi, H.S.; SAE 1990, Paper 900708.
35. Monroe, D.R.; Krueger, M.H.; SAE 1987, Paper 872130.
36. Skowron, J.F.; Williamson, W.B.; Summers, J.C.; SAE 1989, Paper 892093.

RECEIVED February 26, 1992

Chapter 4

Experimental Comparison Among Hydrocarbons and Oxygenated Compounds for their Elimination by Three-Way Automotive Catalysts

J. M. Bart[1], A. Pentenero[2], and M. F. Prigent[1,3]

[1]Institut Français du Pétrole, 92506 Rueil-Malmaison, France
[2]Laboratoire de Chimie-Physique, Université de Nancy I, 54506 Vandoeuvre-les-Nancy, France

Many hydrocarbon species are present in automotive exhaust gases, and three-way Pt-Rh catalysts are commonly used for their elimination. However, most published work on individual hydrocarbon conversion concerns their oxidation in simulated exhaust gases with excess oxygen. This study was therefore undertaken to determine the reactivity of saturated alkanes, olefins, acetylene, aromatics, alcohols or various other oxygenated compounds in steady state conditions with synthetic exhaust gases near stoichiometry. In a first series of measurements, conversion rates were determined as a function of temperature at stoichiometry. The partial pressure effect of O_2, NO and H_2O was then determined at constant temperature in the region of catalyst light-off. NO and mainly O_2 were shown to have a negative effect on the first terms of saturated alkane conversion under lean conditions. Water vapor has a positive effect in rich conditions (without SO_2), but is more pronounced for Pt-Rh than for a Pt catalyst. Finally, the role played by SO_2 in hydrocarbon conversion was evaluated. Its action is sometimes positive, i.e. for saturated alkanes with up to 3 carbon atoms, but more often negative, i.e. in stoichiometric and lean conditions for acetylene and for hydrocarbons for which oxidation starts below 200°C (olefins, aromatics and alkanes with more than 3 carbon atoms), and in rich conditions for all kinds of hydrocarbons.

Several hundred different hydrocarbons and oxygenated compounds can be detected in engine exhaust gases (*1-5*). Some come directly from the fuel. Others, initially not present, are synthesized from hydrocarbon fragments during the combustion and/or exhaust strokes. Their nature varies with the fuel composition, the engine type and its running conditions. Gas chromatographic analysis of exhaust gases shows that they usually contain (in ppm C) around 30% paraffins, 30% olefins of which one half is ethylene, 30% aromatics and 10% acetylene. Depending on their nature, the different hydrocarbons are more or less easily eliminated by automotive catalysts. The only published work on catalyst action on individual

---

[3]Corresponding author

0097–6156/92/0495–0042$06.00/0

hydrocarbons refers to oxidation catalysts used with excess oxygen (*6-10*). No extensive study has been published on the reactivity of individual hydrocarbons on three-way catalysts around stoichiometry.

Concerning alkane oxidation with excess oxygen, it is generally assumed that their adsorption is the limiting step in the reaction. A C-H bond has to be broken, and the more energy involved in this cleavage, the more difficult the oxidation will be. Since the C-H bond is weaker for a secondary carbon atom than for a primary carbon atom, it has actually been observed that propane is easier to oxidize than ethane.

It is also assumed that the number of secondary carbon atoms does not greatly influence alkane adsorption. The introduction of a ternary carbon atom in the hydrocarbon chain on the other hand facilitates the C-H bond cleavage. Branched alkanes are consequently oxidized more easily than hydrocarbons with a straight chain. Isobutane for example was found to oxidize more rapidly than *n*-butane (*10*).

Methane elimination, which is the most difficult to achieve, has been extensively studied on oxidation catalysts (*7*). Formaldehyde can be observed as an intermediate. By the cleavage of one or two C-H bonds, methane is chemisorbed on noble metals as methyl or methylene radicals. These radicals react with an adsorbed oxygenation to give a CH_2O radical and then CO and CO_2.

Alcohol chemisorption is similar to alkane chemisorption and involves the cleavage of a C-H bond (*10*). Ethanol can yield by-products such as ethylene, acetaldehyde and acetone by the following reactions according to Yu Yao (*6*):

$$
\begin{array}{rcl}
C_2H_5OH + 3O_2 & \rightarrow & 2CO_2 + 3H_2O \\
C_2H_5OH + 1/2\,O_2 & \rightarrow & CH_3CHO + H_2O \\
C_2H_5OH & \rightarrow & C_2H_4 + H_2O \\
C_2H_5OH + 2O_2 & \rightarrow & 2CO + 3H_2O \\
2C_2H_5OH + 3/2\,O_2 & \rightarrow & CH_3COCH_3 + CO + 3H_2O
\end{array}
$$

Alkene, like alcohol, may also produce some aldehydes by oxidation (*10*). But in contrast to alkanes, the length of the carbon chain seems to be without any influence on the oxidation rate: 1-hexene and propylene are oxidized with a similar rate (*6*). For ketones also, the chain length has no influence since chemisorption is caused by the carbonyl group.

Aside from the nature of the hydrocarbon species, their oxidation is influenced by many other parameters such as:

- The partial pressure of oxygen, nitrogen oxide, water vapor and sulfur dioxide,
- The nature, concentration and dispersion of the noble metal used (Pt, Pd or Rh),
- The nature of the noble metal carrier (Al_2O_3, CeO_2, etc.),
- Catalyst contamination by poisons.

For three-way catalysts the oxygen partial pressure is rather low and can become substoichiometric if there are momentary excursions of the A/F ratio into the rich region. The water-gas shift reaction,

$$CO + H_2O \rightarrow H_2 + CO_2$$

and the steam-reforming reaction,

$$HC + H_2O \rightarrow CO + CO_2 + H_2$$

can also play a role in CO and HC removal (11).

The role played by SO_2 in hydrocarbon elimination on Pt-Rh three-way catalysts is also very important, and several previous studies have dealt with C_3H_6 or C_3H_8 oxidation in lean or rich conditions (12-16).

An inhibition of propylene oxidation was observed while an unexpected beneficial effect of SO_2 was noted in the oxidation of propane by H.C. Yao et al. The latter was attributed to alumina surface sulfatation, which is able to strongly promote C_3H_8 chemisorption. It was then judged plausible that the new active sites for propane oxidation are located at the junction between Pt particles and the sulfated alumina carrier (14).

Under reducing conditions, SO_2 can be reduced to the S^{2-} sulfide ion, which is a well-known noble metal poison. Very little information is available concerning its effect on hydrocarbon elimination, except that their vaporeforming becomes very difficult in the presence of SO_2 (16).

Given the above situation, a laboratory research program was undertaken to measure the reactivity on three-way catalysts of the main hydrocarbons and more generally of the organic substances present in engine exhaust gases. Their oxidation rate was first determined in the steady state as a function of temperature with a stoichiometric gas mixture. The partial pressure effect of O_2, NO, H_2O and SO_2 was then measured at constant temperature in the region of catalyst light-off.

Experimental Laboratory Methods

The laboratory experimental setup used for measuring hydrocarbon conversion rates is shown in Figure 1. It comprises the gas mixture generating section, the reactor and the analytical equipment. The gas flows are controlled by means of mass flowmeters. Liquid hydrocarbons and water are introduced by metering pumps and vaporized before being mixed with the main gas flow. SO_2 was introduced in the gas mixture for some tests by dissolving it in the injected water.

The gas mixture compositions used for the tests at and around stoichiometry are indicated in Table 1.

Table 1 - Feed gas composition used in the tests (% vol)

CO	0.61	SO_2	0 or 0.002
$HC(C_1)$	0.15	CO_2	10
H_2	0.2	H_2O	10
NO	0.048	N_2	bal
O_2	~0.63		

The oxygen content was adjusted according to the H/C hydrocarbon ratio to obtain the desired redox ratio.

The reactor consisted of a quartz tube (Φ_i = 10 mm) filled, upstream from the catalyst, with quartz scraps to help with inlet gas preheating. The gas flow was 6.67 standard liters per minute giving a gas space velocity in the catalyst of 50,000 h^{-1}. A thermocouple located 5 mm in front of the monolith inlet side was used for temperature determination.

The catalyst used was a 46 mm long cylindrical section of a cordierite monolith (400 cells/in^2), with an alumina based wash-coat (100 g/l) containing 4.5% CeO_2, promoters and 1.41 g/l (40 g/CF) Pt + Rh. The Pt and Rh weight ratio was 5:1.

The results reported here were obtained with the catalyst in fresh conditions (activated 2 hr at 500°C in a stoichiometric mixture), but other experiments with thermally aged catalyst samples were also performed and their main conclusions will be given. To assess if some of the effects observed with a change in the oxygen concentration are more or less linked to the tendency of Rh to be oxidized, some tests were also performed with a Pt catalyst alone.

CO and NO conversions were measured using conventional nondispersive infrared and chemiluminescence analyzers, respectively. Hydrocarbon conversions were measured with a heated flame ionization detector analyzer.

The measurements were made in the steady state, and conversions are defined as the fractional reduction of the inlet concentration, i.e. $X_i = \dfrac{c_i^\circ - c_i}{c_i^\circ} \times 100$

where C_i and C_i° denote outlet and inlet concentrations of component i respectively.

The outlet concentration refers only to the signal magnitude given by the FID analyzer irrespective of the nature of the compound present in the gas. The catalyst light-off temperature is defined as the minimum temperature required to have 50% conversion of the compound considered.

Results and Discussion

HC Conversion Rate at Stoichiometry as a Function of Temperature with or without SO$_2$ present in the gas mixture.

Saturated hydrocarbons. Methane, ethane, propane, and hexane conversions in a stoichiometric gas mixture containing or not 20 ppm SO$_2$ are given as a function of temperature in Figure 2. The light-off temperature T50 decreases very rapidly as the number of molecular carbon atoms increases when starting from methane and levels off above 5 or 6 carbon atoms as shown in Figure 3. A strong positive effect of SO$_2$ is observed for ethane (T50 lowered by 120°C) and propane (T50 lowered by 60°C), and a moderate negative effect is observed for hexane (T50 increased by 10°C).

Olefinic hydrocarbons. Ethylene and propylene conversions in a stoichiometric gas mixture containing or not 20 ppm SO$_2$ are given as a function of temperature in Figure 4. The light-off temperature is 215°C lower for ethylene than for ethane and 100°C lower for propylene than for propane (Figure 3). Conversions for both are 100% above 250°C. A moderate negative effect of SO$_2$ is observed for ethylene and propylene (T50 respectively increased by 10 and 33°C).

Acetylene. Acetylene, which is present in relatively large quantities in exhaust gases, is an easier compound to oxidize than ethane but is less reactive than ethylene. Its light-off temperature (T50) is 285°C in an SO$_2$-free gas mixture and 338°C with 20 ppm SO$_2$ (Figure 5).

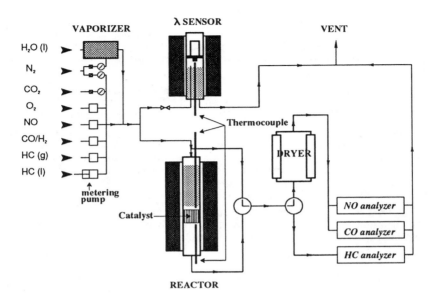

Figure 1.   Schematic diagram of the laboratory test apparatus.

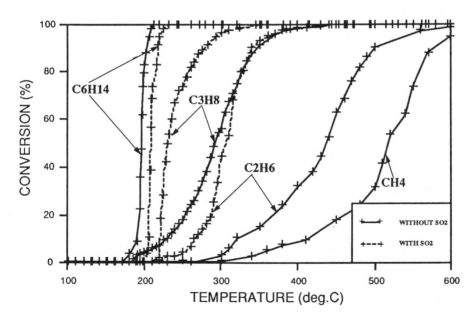

Figure 2.   Alkane conversion rate on Pt-Rh at stoichiometry as a function of temperature
             with and without 20 vpm SO_2.

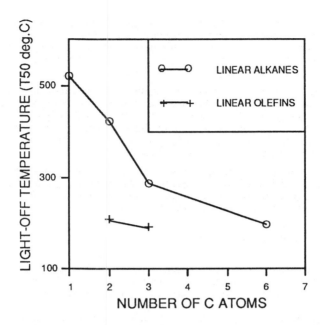

Figure 3.   Light-off temperature in a SO_2-free stoichiometric gas mixture for linear alkanes and olefins as a function of the molecule's number of carbon atoms.

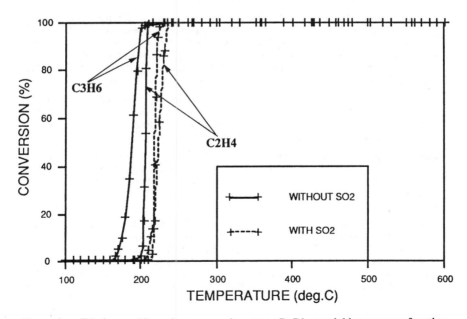

Figure 4.   Ethylene and Propylene conversion rate on Pt-Rh at stoichiometry as a function of temperature with and without 20 vpm SO_2.

Aromatics. Benzene, toluene and orthoxylene conversion in a stoichiometric gas mixture are given as a function of temperature in Figure 6. They behave very similarly to olefins with light-off temperatures (T50) around 200°C for the three species tested. The effect of SO_2 was measured only for benzene and is very moderate (T50 increased by 14°C).

Alcohols. Methanol, ethanol, *n*-propanol, *iso*-propanol and *n*-butanol-1 conversion in an SO_2-free stoichiometric gas mixture are given as a function of temperature in Figure 7. They beha0ve very similarly with light-off temperatures around 200°C, like olefins. In contrast to alkanes, their conversion rates are not very dependent on the number of molecular carbon atoms. Methanol is as easy to oxidize in these conditions, as are heavier alcohols.

Other oxygenated compounds. Methyltertiarybutylether (MTBE) and Ethyltertiarybutylether (ETBE), frequently added to unleaded gasoline as octane boosters, and acetone were also evaluated. Their reactivity on Pt-Rh TWC is very similar to the reactivity of alcohols as mentioned above.

Effect of Oxygen Partial Pressure

In engine exhaust gases, frequent deviations from stoichiometry occur. It is then very important to know how the hydrocarbon conversion is affected by lean or rich excursions. The following results were obtained in the steady state at constant temperature by changing, step-by-step, the oxygen concentration between zero and about twice the quantity corresponding to stoichiometry. All other gas concentrations remained constant, with 20 ppm SO_2 being added or not to the gas mixture. The operating temperature was usually low and chosen to be just above the hydrocarbon light-off temperature (usually 260°C, but 190°C and 320°C or 380°C were also used for hydrocarbons easy or difficult to oxidize).

Saturated Hydrocarbons. Ethane, propane, pentane and hexane conversions are plotted as a function of the test gas O_2 concentration respectively in Figures 8 and 9 at temperatures of 190, 260 or 380°C. Their conversion increases when the O_2 concentration is raised to values around or below stoichiometry and then decreases either sharply (ethane and propane) or moderately (pentane and hexane) when the O_2 concentration goes up to values exceeding stoichiometry.

Figure 10 shows that for propane and hexane the same type of conversion curves are obtained at 260°C with the Pt catalyst as with the Pt-Rh catalyst. It can then be assumed that the drop in conversion observed in the lean region is not a direct consequence of the Pt crystallite coverage by Rh oxide.

This negative effect of an excess of oxygen was already observed for alkane oxidation on Pt by YuYao (6). This has been explained, assuming that the slow reaction step is the dissociative alkane chemisorption, with the breakage of the weakest C-H bond, followed by interaction with an adsorbed oxygen atom on an adjacent site (7,10).

Oxygen and alkane, with the latter taking up probably several adjacent sites, are then in competition for the surface coverage leading consequently to an inhibition by O_2 when in excess as found for ethane or propane and to a lesser extent for pentane or hexane. For long-chain hydrocarbons, which are easily chemisorbed and react readily, it was expected that the Pt surface is in a more reduced state for the same O_2 partial pressure, and should then be less sensitive to an excess of oxygen.

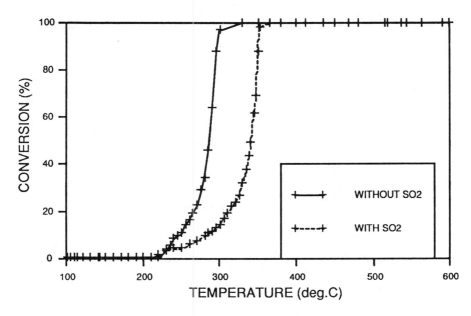

Figure 5. Acetylene conversion rate on Pt-Rh at stoichiometry as a function of temperature with and without 20 vpm SO_2.

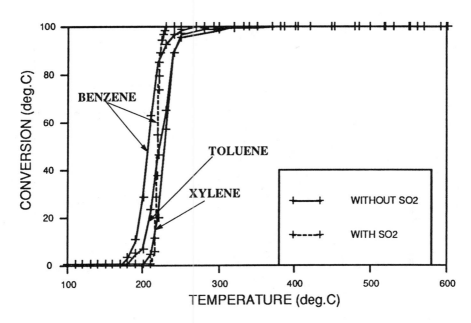

Figure 6. Benzene, Toluene and o-Xylene conversion rate on Pt-Rh at stoichiometry as a function of temperature with and without 20 vpm SO_2.

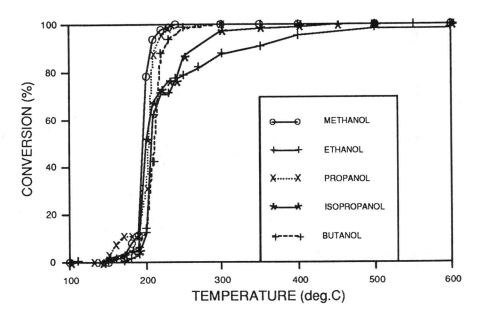

Figure 7.   Alcohol conversion rate on Pt-Rh at stoichiometry as a function of temperature (without SO_2).

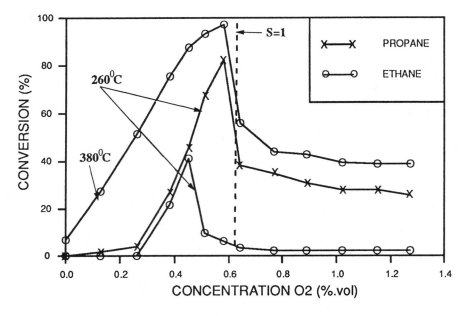

Figure 8.   Ethane and Propane conversion rate on Pt-Rh at 260 and 380°C as a function of O_2 concentration (without SO_2) below and above stoichiometry (S = 1).

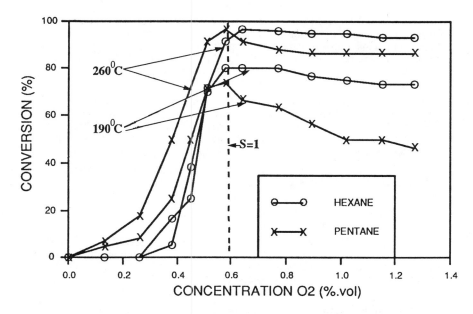

Figure 9. Pentane and Hexane conversion rate on Pt-Rh at 190 and 260°C as a function of O_2 concentration (without SO_2) below and above stoichiometry (S = 1).

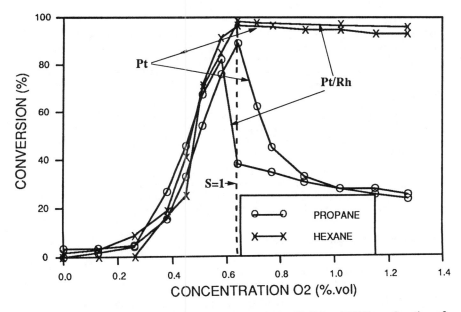

Figure 10. Propane and Hexane conversion rate on Pt and Pt-Rh at 260°C as a function of O_2 concentration (without SO_2) below and above stoichiometry (S = 1).

The addition of 20 ppm SO_2 has an effect on propane conversion which is slightly negative in the rich region (for $O_2 < 0.5\%$), but strongly positive in the lean region (Figure 11). This is probably the consequence, as said previously, of the possibility of alkane adsorption on sulfated alumina, and especially around the Pt crystallites, thus opening up the possibility of a reaction between oxygen atoms adsorbed on Pt sites and these alkane molecules chemisorbed on the sulfated alumina (14).

The conversion of hexane as a function of temperature in a rich gas mixture (O_2 = 0.49%) occurs in two steps : a low temperature oxidation by O_2 followed by a higher temperature vaporeforming by H_2O (Figure 12). In the presence of 20 ppm SO_2, only the low temperature step occurs, with the vaporeforming reaction then being completely inhibited by SO_2 or by a sulfur compound with a lower oxidation state.

Olefins. Ethylene and propylene conversions on the Pt-Rh catalyst are plotted as a function of the test gas O_2 concentration in **Figure** 13 at the temperature of 260°C.

For both, conversion increases to stoichiometry and then remains close to 100% with further O_2 concentration increases.

Acetylene. Figure 14 shows that for acetylene, at 320°C, a net excess of oxygen is necessary, since its conversion does not reach a maximum at stoichiometry: a 50 % oxygen excess is required to reach 100 % conversion. Higher O_2 concentrations do not decrease acetylene conversion. SO_2 has a negative effect in both the rich and lean regions up to about twice the stoichiometric O_2 level.

Aromatics. Figure 15 shows that benzene and o-xylene behave like olefins with conversion increasing up to stoichiometry and remaining constant close to 100% with higher O_2 levels.

Effect of NO Partial Pressure

With saturated aliphatic hydrocarbons, a conversion drop was observed with O_2 concentrations above stoichiometry. Among other hypotheses, this drop could be attributed to an inhibition by NO which remains adsorbed on the catalyst surface because its reduction does not occur in the presence of excess O_2. To check the validity of this hypotheses similar tests were performed with and without NO in the test gas.

Figure 16 shows that, by comparison with the results shown previously in Figure 10, NO has a negative effect on propane conversion on Pt-Rh or Pt catalysts. But even without NO a sharp drop in propane conversion is still observed above stoichiometry.

Figures 17 and 18 quantify this NO effect for various levels of O_2 concentration in the test gas, for the Pt and Pt-Rh catalysts at a temperature of 260°C. The NO effect is greatest just above stoichiometry for the Pt-Rh catalyst. It remains significant for higher O_2 levels, notably higher than stoichiometry for the Pt catalyst.

Effect of H₂O Vapor Pressure

Water vapor is assumed to play a role in hydrocarbon elimination by reforming reactions when the O_2 concentration becomes substoichiometric.

Figure 16 shows that in steady state conditions propane conversion at 260°C is actually affected by H_2O on the Pt-Rh catalyst and, to a lesser extent, on Pt.

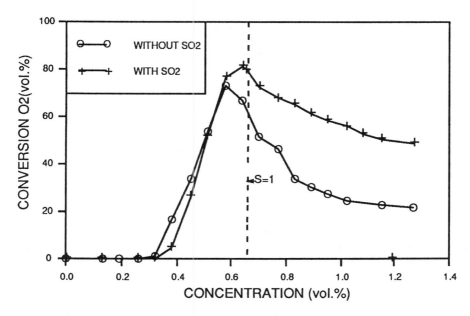

Figure 11. 20 vpm SO_2 effect on propane conversion rate on Pt-Rh at 260°C as a function of O_2 concentration below and above stoichiometry (S = 1).

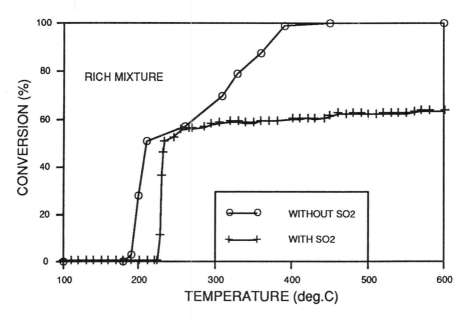

Figure 12. 20 vpm SO_2 effect on hexane conversion rate on Pt-Rh in a rich mixture (O_2 = 0.49%) as a function of temperature.

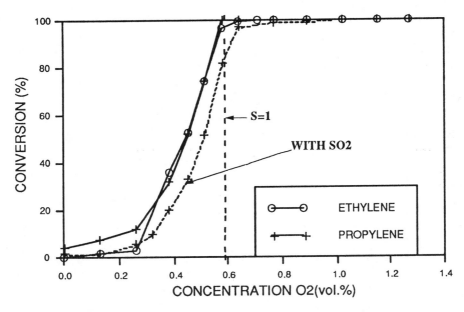

Figure 13. Ethylene and Propylene conversion rate on Pt-Rh at 260°C as a function of O_2 concentration, below and above stoichiometry (S = 1), with and without 20 vmp SO_2.

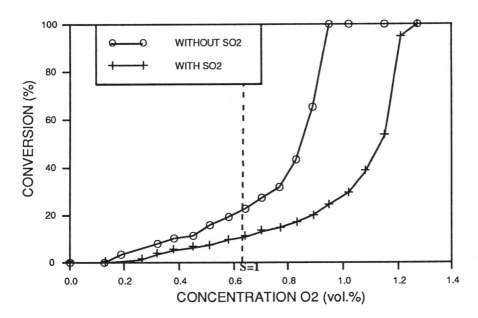

Figure 14. Acetylene conversion rate on Pt-Rh at 320°C as a function of O_2 concentration, below and above stoichiometry (S = 1), with and without 20 vpm SO_2.

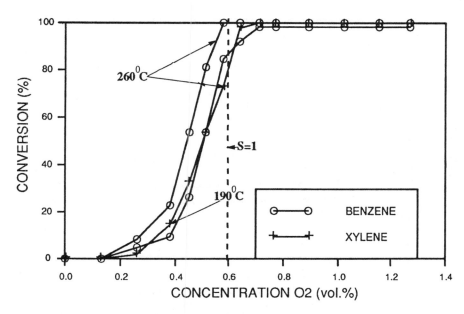

Figure 15. Benzene and o-Xylene conversion rate on Pt-Rh at 190 and 260°C as a function of O_2 concentration (without SO_2) below and above stoichiometry (S = 1).

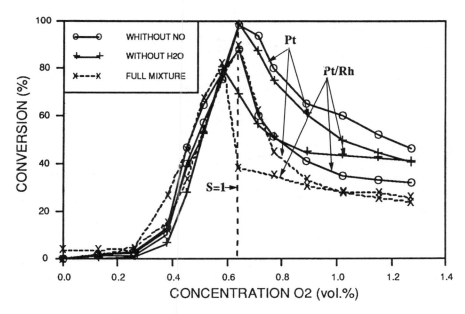

Figure 16. Effect of NO (480 vpm) and H_2O (10%) on Propane conversion rate on Pt and Pt-Rh at 260°C (without SO_2), as a function of O_2 concentration below and above stoichiometry (S = 1).

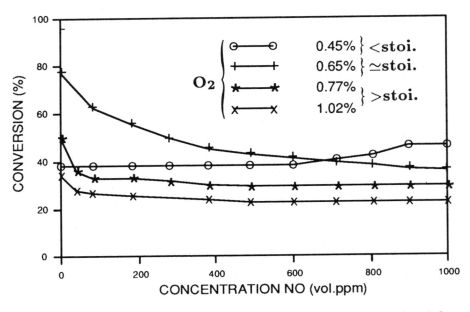

Figure 17. Propane conversion rate on Pt-Rh at 260°C as a function of NO and O_2 concentrations.

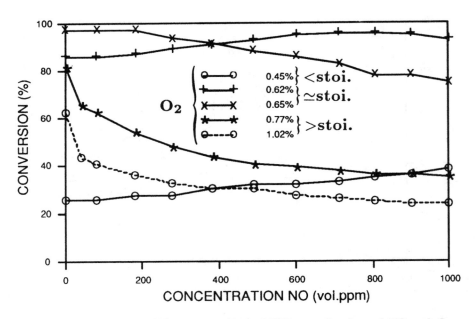

Figure 18. Propane conversion rate on Pt at 260°C as a function of NO and O_2 concentrations.

Figures 19 and 20 quantify this H_2O effect for various levels of O_2 concentrations at 260°C. The effect on conversion is positive with O_2 concentrations below stoichiometry and negative above. The positive water vapor effect is however completely eliminated by SO_2 as has already be shown for hexane.

Effect of catalyst aging

A similar Pt-Rh catalyst sample was aged at 900°C in N_2 + 10% H_2O during 16 hours. The resulting increase in noble-metal particle size was measured by CTEM. From about 6 nm in the new catalyst it rose to 20 nm in the aged catalyst (with a simultaneous rearrangement of Pt and Rh on the metal crystallites observable by STEM but difficult to quantify).

Several of the previous tests were performed again with this aged catalyst. At stoichiometry an increase in the light-off temperatures between 15°C (for benzene) and 100°C (for propane) was noticed. The hydrocarbons classification according to their light-off temperature were modified as follows:

New catalyst $C_3H_6 \geq C_6H_{14} > C_2H_4 \approx C_6H_6 >> C_3H_8 >> C_2H_6$
Aged catalyst $C_6H_6 \geq C_6H_{14} \geq C_3H_6 \geq C_2H_4 >> C_3H_8 >> C_2H_6$

Concerning hydrocarbon conversion as a function of the gas mixture oxygen content and, in the absence of SO_2, the following evolutions due to catalyst aging were observed:

- For propane, the conversion rate decrease was the most pronounced below stoichiometry where the conversion spike observed (cf. Figure 8) was lowered from 80% down to 40%,
- A moderate inhibiting effect of NO on propane was observed as with the new catalyst (cf. Figure 16),
- A moderate inhibiting effect of O_2, not existing with the fresh catalyst (cf. Figure 6) was noticed at 260°C in the lean region with benzene,
- Hexane vaporeforming (cf. Figure 12) was still possible, but was shifted 40 to 50°C towards higher temperatures.

Summary and Conclusions

The experimental results given above show that, on a Pt-Rh three-way catalyst working at stoichiometry,

- Saturated hydrocarbon conversion is dependent on the number of carbon atoms present in the molecule, but that this effect is noticeable only for the first five atoms.
- Olefins have light-off temperatures lower than alkanes having the same number of carbon atoms.
- Acetylene is easier to oxidize than ethane but is less reactive than ethylene.
- Aromatics such as benzene, toluene or xylene behave very similarly and have light-off temperatures comparable to olefins.

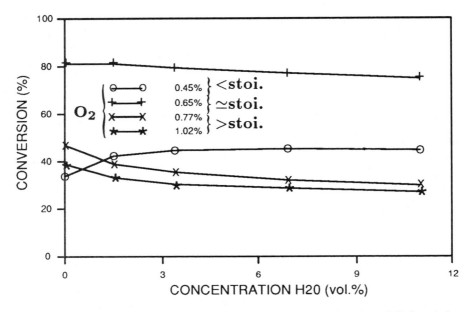

Figure 19. Propane conversion rate on Pt-Rh at 260°C as a function of H_2O and O_2 concentrations.

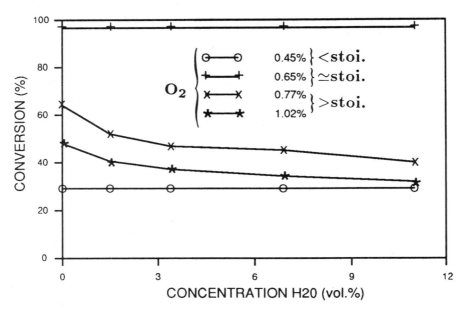

Figure 20. Propane conversion rate on Pt at 260°C as a function of H_2O and O_2 concentrations.

- Alcohols and other oxygenated compounds (MTBE, ETBE or acetone) have conversion rates nearly independent of the number of carbon atoms present in the molecule, with methanol being converted as easily as the higher terms.

It has also been shown that, when the gas mixture composition deviates from stoichiometry, the lack or excess of oxygen has different effects depending on the nature of the hydrocarbon considered.

- Saturated hydrocarbons have a conversion that first increases with the O_2 concentration up to about stoichiometry, and then decreases either sharply (ethane and propane) or moderately (pentane and hexane)

- Olefins and aromatics have a conversion that first increases with the O_2 concentration up to stoichiometry and then remains constant close to 100% without any drop with further O_2 concentration increases.

- Acetylene has a conversion that increases with O_2 concentration below and above stoichiometry, and a 50% O_2 excess is necessary to reach 100% C_2H_4 conversion.

Nitric oxide has a negative effect on propane conversion for O_2 levels above stoichiometry, but this effect is not enough to explain the C_3H_8 conversion drop observed with lean mixtures because this occurs even without NO present.

Water vapor also affects HC conversion. The effect is more pronounced with a Pt-Rh than with a Pt catalyst. For propane the effect is positive with O_2 concentrations below stoichiometry and negative above stoichiometry.

A very important role is also played by SO_2 in hydrocarbon conversion, which can be summarized as follows.

- In lean and stoichiometric exhaust gases
  - positive effect of SO_2 for saturated aliphatic hydrocarbons with a number of carbon atoms up to 3;
  - negative effect of SO_2 for acetylene and for hydrocarbons whose oxidation starts below 200°C (olefins, aromatics, and saturated aliphatic HC with a number of carbon atoms above 3).

- In rich exhaust gases
  - negative effect on catalyst light-off for all hydrocarbons
  - inhibition of the vaporeforming reaction (observed for n hexane).

Acknowledgment

The authors are grateful to Octel S.A., PSA and RNUR for financial support for this project.

REFERENCE

(1)   J.K. Walker and C.L. O'Hara: "Analysis of Automobile Exhaust Gases by mass Spectrometry". Anal. Chem. 27 (5), 825-828, 1955.

(2)   E.S. Jacobs: "Rapid Gas Chromatographic Determination of C_1 to C_{10} Hydrocarbons in Automotive Exhaust gas". Anal. Chem. 38 (1), 43-47, 1966.

(3)   D.J. McEven: "Automobile Exhaust Hydrocarbon Analysis Gas Chromatography". Anal. Chem. 38 (8), 1047-1053, 1966.

(4)   M.W. Jackson: "Effect of Catalytic Emission Control on Exhaust Hydrocarbon Composition and Reactivity" SAE paper No 780624, 1978.

(5)   F.M. Black, L.E. High and J.M. Lang: "Composition of Automobile Evaporative and Tailpipe Hydrocarbon Emissions." J.A.P.C.A., 30 (11), 1216-1221, 1980.

(6)   Y.F. Yu Yao: "Oxidation of Alkanes over Noble Metal Catalysts" Ind. Eng. Chem. Prod. Res. Dev. 19, 293-298, 1980.

(7)   C.F.Cullis, D.E. Keene and D.L. Trimm: "Studies of the Partial Oxidation of Methane over Heterogeneous Catalysts" J.Catal. 19, 378-385, 1970.

(8)   E. Koberstein:"Oxidation Catalyst for the reduction of hydrocarbon emissions" VDI Berichte 578, 151-165, 1985.

(9)   M.A. Accomazzo and Ken Nobe: "Catalytic Combustion of C_1 to C_3 Hydrocarbons". Ind. Eng. Chem. Proc. Des. Dev. 4 (4), 425-430, 1965.

(10) A. Schwartz, L.L. Holbrook and H. Wise: "Catalytic Oxidation Studies with Platinum and Palladium", J. Catal. 21, 199-207, 1971.

(11) J.T. Kummer: "Catalyst for Automobile Emission Control", Prog. Energy Comb Sci. 6, 177-179, 1981.

(12) H.C. Yao, H.K. Stepien and H.S. Gandhi: "The effects of SO_2 on the oxidation of hydrocarbons and carbon monoxide over Pt/γ-Al_2O_3 catalyst". J. Catal, 67, 231-236, 1981.

(13) V.I. Panchishnyi, M.K. Bondareva, A.V. Sklyarov, V.V. Rozanov, and G.P. Chadina. "Oxidation of carbon monoxide and hydrocarbons on platinum and palladium catalysts in the presence of sulfur dioxide". Zhurnal Prikladnoi Khimii 61 (5), 1093-1098, 1998.

(14) H.S. Gandhi and M. Shelef: "Effects of sulfur on noble metal automotive catalysts". Applied Catal. 77, 175-186, 1991.

(15) C.C. Chang: "Infrared studies of SO_2 on γ-alumina". J. Catal., 53, 374-385, 1978.

(16) J.C. Summers and K. Baron: "The effects of SO_2 on the performance of noble metal catalysts in automobile exhaust". J. Catal., 57, 380-389, 1979.

RECEIVED February 12, 1992

Chapter 5

Steady-State Isotopic Transient Kinetic Analysis Investigation of CO–O$_2$ and CO–NO Reactions over a Commercial Automotive Catalyst

Rachid Oukaci, Donna G. Blackmond, James G. Goodwin, Jr., and George R. Gallaher[1]

Chemical and Petroleum Engineering Department, University of Pittsburgh, Pittsburgh, PA 15261

Steady-state isotopic transient kinetic analysis (SSITKA) was used to study two model reactions, CO oxidation and CO-NO reactions, on a typical formulation of a three-way auto-catalyst. Under steady-state conditions, abrupt switches in the isotopic composition of CO ($^{12}C^{16}O/^{13}C^{16}O + ^{13}C^{18}O$) were carried out to produce isotopic transients in both labeled reactants and products. Along with the determination of the average surface lifetimes and concentrations of reaction intermediates, an analysis of the transient responses along the carbon reaction pathway indicated that the distribution of active sites for the formation of CO$_2$ was bimodal for both reactions. Furthermore, relatively few surface sites contributed to the overall reaction rate.

Typical automotive catalytic converters contain rhodium and platinum as the primary catalytic materials which can carry out simultaneously carbon monoxide (CO) and hydrocarbon oxidation as well as nitric oxide (NO) reduction. These noble metals, especially rhodium, are in limited supply and very expensive. Significant interest in CO oxidation and CO-NO reaction on Pt/Rh catalysts continues because of the need to utilize better these valuable metals. Understanding the mechanism of the reactions on these metals and their kinetics at a site level may provide ways to develop more efficient or inexpensive catalysts for the reduction of pollutants in automotive exhausts.

The utilization of transient methods in kinetic studies, especially when carried out under steady-state reaction conditions, can provide basic kinetic information which cannot be obtained via traditional steady-state or non-steady state transient experiments. Steady-state isotopic transient kinetic analysis

[1]Current address: Alcoa Separations Technology Division, Warrendale, PA 15086

(SSITKA) has been successfully used to investigate the kinetics of a number of important reactions, including CO hydrogenation (1-7), ammonia synthesis (8), and oxidative coupling of methane (9-10). An analysis of the transient responses allows the determination of the average surface retention time and concentration of reaction intermediates under the unperturbed steady-state reaction conditions. With proper analysis based on certain assumptions, SSITKA permits quantification of the surface site heterogeneity and determination of the relative contribution of the various active sites to the overall reaction rate.

In the present study, SSITKA was used to investigate the kinetics of both CO-O_2 and CO-NO reactions on a typical formulation of a three-way auto-catalyst. This catalyst contained the active metals Pt and Rh supported on Al_2O_3 promoted with CeO_2 and NiO. Under steady-state conditions, abrupt switches in the isotopic composition of CO ($^{12}C^{16}O/^{13}C^{16}O + ^{13}C^{18}O$) were carried out to produce isotopic transients in both labeled reactants and products. Under such conditions, the transient measurements were able to be obtained without perturbing the steady-state reaction environment.

Experimental
The catalyst used for this study was provided by General Motors Co.-AC Rochester Division. This was a typical commercial three-way auto-catalyst formulation used for spraying the monolith substrate. It was obtained in powder form and had the following composition by weight:

Rh	0.069%
Pt	0.967%
NiO	2.0%
CeO_2	28%
Al_2O_3	Balance

This catalyst was calcined at 400-500°C. In order to estimate the active metal site concentration on the catalyst, static hydrogen chemisorption was carried out after reduction in hydrogen. A mono-metallic 3%Rh/La_2O_3 catalyst was also used for comparison purposes.

Both the steady-state and the transient reaction rates were measured in a micro-reactor consisting of a 4 mm I.D. straight quartz tube narrowed to a 1 mm opening after the catalyst bed. The catalyst was held in place by glass wool. Catalyst loadings varied between 25 and 100 mg. The products and reactants were analyzed on-line using a Varian 3700 gas chromatograph equipped with a thermal conductivity detector. The gases were separated by a 60/80 Carbosieve S-II column. An Extrel quadrupolar mass spectrometer interfaced with a microcomputer for data acquisition was used to monitor the evolution and decay of different labeled reactants and products following an isotopic switch.

Ultra-high purity gases from Linde were used for both reactions. The non-labeled carbon monoxide (^{12}CO) contained about 5% argon (Ar) as a tracer to measure the gas phase hold-up during the transient experiments. Helium was used as a diluent.

CO oxidation was carried out under differential conditions at atmospheric pressure and 95-175°C using a reactant mixture of $CO:O_2:He = 2.5:2.5:95$ flowing at 68 cc/min.

NO reduction by CO was also carried out at 1 atm, but higher temperatures were required to obtain a measurable conversion (175-300°C). The composition of the reactant mixture was: $CO:NO:He = 2.5:2.5:95$ and the flow rate was 68 cc/min.

The steady-state isotopic transient experiments were performed at 95°C for CO oxidation and 200°C for CO-NO reaction. With the reaction at steady-state for the conditions of each experimental run, an abrupt switch was made from a reactant feed containing ^{12}CO + Ar to one containing an equivalent amount of ^{13}CO and flowing at the same flow rate (Fig. 1). This procedure insured that the concentrations of all adsorbed species and, therefore, the overall rate of formation of carbon-containing products remained constant. Typical curves (rates versus time) of the evolution and decay of the products (CO_2) are also shown schematically in Figure 1. Similar curves were obtained for CO. The ^{13}CO used in these experiments contained also about 10% $^{13}C^{18}O$ which allowed the acquisition of further information by monitoring the labeled oxygen in the products.

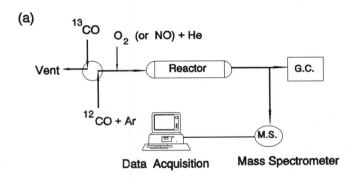

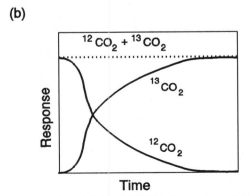

Figure 1.   Schematic of (a) a SSITKA System and (b) Typical Steady-State Transient curves

Results and Discussion

CO Oxidation. The steady state reaction results obtained at 95 °C for the three-way auto-catalyst (referred to as TW) and at 100 °C for the Rh/La_2O_3 catalyst are shown in Table I. Although the Rh/La_2O_3 catalyst had a higher metal loading, it had a much lower activity for CO oxidation. It is well known that Pt is a better catalyst for this reaction. The apparent activation energy for this reaction on the TW catalyst was found to be equal to 14.7 kcal/mol. This value is about half that reported for CO oxidation over supported Rh catalysts (11), but it is in agreement with activation energies reported for this reaction on supported Pt catalysts (12). Thus, it may be reasonably assumed that the bulk of CO oxidation was occurring on the Pt surface.

Table I.   CO Oxidation: Steady-State Reaction Data

	TW Cat.	Rh/La_2O_3
Reaction Temp. [°C]	95	100
CO Conv. [%]	0.62	0.95
Rate [μmol/g_{cat}.s]	0.22	0.08
TOF[#] [s^{-1}]x10^3	6.5	1.0

[#] Based on H_2 chemisorption

Figure 2 shows typical steady-state transient curves for labeled CO and CO_2 and argon obtained after an abrupt switch from (^{12}CO + Ar) to ^{13}CO over the TW catalyst. The curves for ^{13}CO and $^{13}CO_2$ were actually obtained by inverting the curves obtained experimentally in order to compare them with that of Ar. The CO transient indicates either that the CO adsorption/desorption step was very fast or that CO adsorption was essentially irreversible. The CO_2 response was found to significantly lag that of Ar (the gas phase marker) indicating that the intermediate to CO_2 or the CO_2 itself was held on the catalyst surface. Similar transient response curves were obtained for the Rh/La_2O_3 catalyst.

The average surface lifetime (τ) of the ^{13}C-surface intermediates leading to CO_2 was determined from the area between the CO_2 transient and Ar curves (the latter representing the gas phase holdup). The reciprocal of the CO_2-intermediate residence time, $1/\tau$, represents a measure of the reaction rate constant and the intrinsic site turnover frequency (3-4). The surface population or abundance, N_{*CO2}, of these intermediates can be determined from the product of the steady-state reaction rate and the surface lifetime. Assuming that CO_2 did not readsorb under the reaction conditions used, then N_{*CO2} can be considered as a better measure of the actual number of active reaction sites than that measured by hydrogen chemisorption. Reaction of CO_2 with ceria to form carbonates, which is known to occur with most rare earth oxides (13), and their subsequent decomposition during the isotopic switch   may result in an overestimation of N. Similarly, if CO_2 readsorption/desorption occurs, then $N_{\cdot CO2}$ would represent an upper limit of the number of active sites. All these parameters are given in Table II for both the TW catalyst and the Rh/La_2O_3 catalyst with the fractional coverages based on the metal surface area estimated by hydrogen chemisorption.

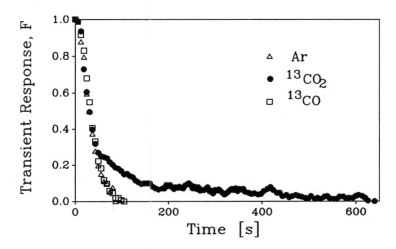

Figure 2. Steady-State CO Oxidation at 95°C: Normalized Isotopic Transient Responses

The surface concentrations of the CO_2 intermediates in Table II suggests that not all the surface measured by hydrogen chemisorption was available for reaction, at least under these reaction conditions. The results were even more dramatic for the Rh/La_2O_3 catalyst for which only 5% of the exposed metal atoms measured by hydrogen chemisorption participated directly in CO_2 formation.   The remainder of the surface may have been occupied by nonreactive carbon species or by other ad-species participating in the reaction such as oxygen.

The results in Table II indicate also that the average site activity for both catalysts was much higher than the TOF which was based, as is done commonly, on the number of sites determined by hydrogen chemisorption .

Further analysis of the steady-state transient CO_2 responses was carried out to observe any kinetic manifestation of surface heterogeneity.   The semilogarithmic representation of the transient rate data was not linear but concave upward, indicating a distribution of retention times for CO_2 precursors.

Table II. Steady-State CO Oxidation: Estimation of Surface and Kinetic Parameters

	TW Cat.	Rh/La_2O_3
Reaction Temperature [°C]	95	100
Surface Lifetime, τ [s]	55.3	47.2
Average Site Activity, $k = 1/\tau$ [s^{-1}]x10^3	18.1	21.2
Abundance, $N_{\cdot CO_2}$ [μmol/g$_{cat}$]	11.9	3.7
Surface Coverage[#], θ	0.36	0.05
Activity of site pool 1, k_1 [s^{-1}]x10^3	7.9	7.3
Activity of site pool 2, k_2 [s^{-1}]x10^3	64.7	124.9
Activity Distribution, x_1/x_2	0.70/0.30	0.90/0.10
Rel. Contribution to React. Rate, y_1/y_2	0.22/0.78	0.68/0.31

[#] Based on H_2 chemisorption

Such a distribution of τ's points to a nonuniformity of the active surface of the catalyst (8). Figure 3 illustrates the activity distribution over the active surface of the TW catalyst using the Laplace transform method developed by de Pontes et al. (6) for the deconvolution of isotopic transient response curves. It is clear from Figure 3 that the active sites had not only a wide range of kinetic strengths but also that their activity distribution was bimodal. Most of the active sites had very low activities comparable to the TOF determined in the traditional manner, while 30% of the sites had an average activity which was an order of magnitude higher. A similar bimodal activity distribution was also obtained with the Rh/La_2O_3 catalyst, suggesting the presence of two types of active sites. The average site activities k_1 and k_2 corresponding to the low and high activity pools, respectively, for both catalysts are summarized in Table II.

The product of the activity distribution function f(k) by the site activity k determines the contribution of a pool of active sites of any particular strength to the overall rate of reaction. This is illustrated in Figure 4 for the TW catalyst as the relative contributions of the active sites to the reaction rate as a function of the site activity, k. These results indicate that more than 75% of the activity of the TW catalyst could be attributed to relatively few high activity sites (30%). They are compared to the results obtained for the Rh/La_2O_3 catalyst in Table II. Since this catalyst was not very active for CO oxidation, most of the activity was associated with the low activity sites which comprised about 90% of the total active sites.

The first interpretation of the bimodal distribution in the kinetic strength of the sites on the TW catalyst which comes to mind is the presence of two metals in this catalyst, Pt and Rh, which are known to have different activities for CO oxidation. However, the observation of a similar bimodal distribution on the mono-metallic Rh/La_2O_3 catalyst indicates that this phenomenon cannot be attributed only to the bimetallic nature of the TW catalyst. Although direct CO

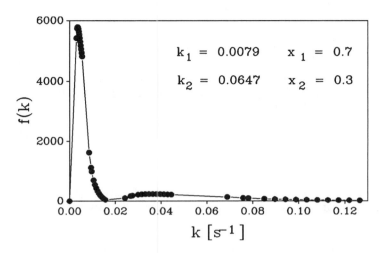

Figure 3. Steady-State CO Oxidation at 95°C: Activity Distribution of the Catalytic Sites (k_i = average site pool activity [s^{-1}], x_i = fraction of sites of activity k_i)

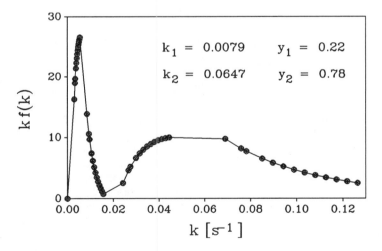

Figure 4. Steady-State CO Oxidation at 95°C: Reactivity Contribution of the Active Sites to the Reaction Rate (k_i = average site pool activity [s^{-1}], y_i = relative contribution of the pool of sites of activity k_i)

oxidation on the support or on the ceria in the case of the TW catalyst may be ruled out, CO_2 may react with the rare earth oxide to form carbonates. A subsequent decomposition of these carbonates could give rise to a wide distribution in k_i and an apparent low activity pool of reaction sites. The low activity pool of sites on both catalysts may be due to multiple desorption-readsorption processes on the metal sites. Such a process could explain why the low activity reaction pool (k_1) was higher for the Rh/La_2O_3 catalyst which had a 2.5 times higher metal surface area per gram catalyst than the TW catalyst. The apparent contribution of these low activity sites to the overall activity was also much higher on the mono-metallic catalyst. Finally, the bimodal site distribution may have resulted from the presence of two types of active metal sites with different kinetic strengths.

CO-NO Reaction. The steady state reaction results at 200°C are shown in Table III. The apparent activation energy for this reaction on the TW catalyst was found to be equal to 10.8 kcal/mol which is only slightly lower than that reported for CO_2 formation from CO-NO reaction over a supported Rh catalyst (11).

Table III.   CO-NO Reaction: Steady-State Reaction Data at 200°C

	TW Cat.	Rh/La_2O_3
CO Conv. [%]	2.08	4.55
Rate [$\mu mol/g_{cat}.s$]	0.73	0.44
TOF[#] [s^{-1}]x10^3	21.9	5.3

[#] Based on H_2 chemisorption

Figure 5 shows the transient curves for labeled CO and CO_2 and Ar obtained after the switch from (^{12}CO + Ar) to ^{13}CO for the TW catalyst. Again, the curves for ^{13}CO and $^{13}CO_2$ were actually obtained by inverting the curves obtained experimentally in order to integrate for surface lifetimes, τ. The average surface lifetimes of the CO_2 intermediates, their abundances, and the average site activities are given in Table IV for both the TW and the Rh/La_2O_3 catalysts. The fractional coverages of active CO_2 intermediates based on the active metal surface area estimated by hydrogen chemisorption are also presented.

The CO_2-precursor surface coverage under these reaction conditions was slightly higher than unity for the TW catalyst. The coverage obtained for the Rh/La_2O_3 catalyst, though low compared to the overall metal surface area measured by hydrogen chemisorption, was much higher than the one previously observed during CO oxidation. Such high coverages may be attributed to several causes. Parts of the support may be contributing adsorption/desorption sites at the high reaction temperature in addition to the metal sites measured by hydrogen chemisorption at ambient temperature. In addition, as indicated in the previous section, reaction of CO_2 with ceria to form carbonates, which is known to occur with most rare earth oxides (13) and their subsequent decomposition during the isotopic switch may also result in an overestimation of N and θ. Similarly, readsorption phenomena on the metal would lead to higher values for the average residence time, τ. Finally, hydrogen chemisorption may not be the most accurate method for estimating the number of active sites.

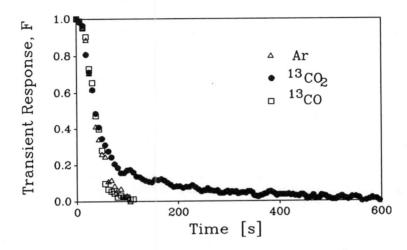

Figure 5. Steady-State CO-NO Reaction at 200°C: Normalized Transient Responses

Table IV. Steady-State CO-NO Reaction at 200°C: Estimation of Surface and Kinetic Parameters

	TW Cat.	Rh/La$_2$O$_3$
Surface Lifetime, τ [s]	53.0	60.5
Average Site Activity, $k=1/\tau$ [s^{-1}]x10^3	18.9	16.5
Abundance, $N_{\cdot CO_2}$ [μmol/g$_{cat}$]	38.6	26.4
Surface Coverage[#], θ	1.16	0.32
Activity of site pool 1, k_1 [s^{-1}]x10^3	4.9	5.1
Activity of site pool 2, k_2 [s^{-1}]x10^3	52.9	43.6
Activity Distribution, x_1/x_2	0.70/0.30	0.76/0.24
Rel. Contribution to React. Rate, y_1/y_2	0.25/0.75	0.60/0.40

[#] Based on H$_2$ chemisorption

Figure 6 shows the site activity distribution on the TW catalyst while Figure 7 illustrates the relative contributions of the active sites to the reaction rate. These results as well as those for Rh/La_2O_3 in Table IV indicate that the catalyst surfaces were not homogeneous for this reaction, and that the distribution of active sites for the formation of CO_2 from CO and NO was bimodal on both the TW catalyst and the mono-metallic Rh/La_2O_3 catalyst. Again, more than 75% of the activity could be attributed to only 30% of the active surface sites on the TW catalyst while on the Rh/La_2O_3 catalyst it was the low activity sites (76%) which seemed to contribute most to the overall reaction rate. As indicated previously, the low activity reaction pool could have possibly been due to multiple desorption-readsorption processes on metal sites. However, neither the presence of two distinct types of active metal sites nor reaction with the rare earth oxides composing the support can be ruled out completely without further experiments which are required to clarify these phenomena.

The residence times obtained for CO_2 from both reactions on the TW catalyst were about the same although the temperature for NO-CO reaction was twice that of CO oxidation. The high τ for the former reaction at the higher temperature indicates that the slow step in this reaction is different from the controlling step in CO oxidation. This result is in agreement with previous suggestions that the slowest step in CO-NO reaction is the dissociation of adsorbed NO, at least on supported Rh catalysts (*11,14*).

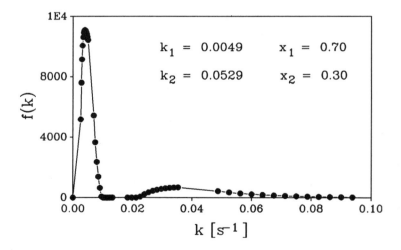

Figure 6. Steady-State CO-NO Reaction at 200°C: Activity Distribution of the Active Sites (k_i = average site pool activity [s⁻¹], x_i = fraction of sites of activity k_i)

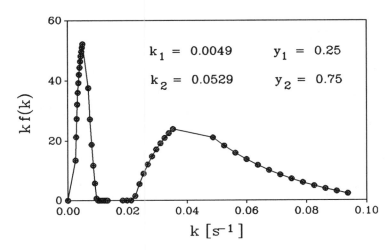

Figure 7. CO-NO Reaction at 200°C: Reactivity Contribution of the Active Sites to the Reaction Rate (k_i = average site pool activity [s^{-1}], y_i = relative contribution of the pool of sites of activities k_i)

An analysis of the transient responses along the ^{18}O-traced trajectory, under both CO oxidation and CO-NO reaction conditions, revealed the formation of both $^{12}C^{18}O_2$ and the scrambled $^{12}C^{16}O^{18}O$. These products suggest that dissociation of CO can possibly occur as suggested by Cho and Stock (*15*). Reaction of the CO_2 with the support accompanied by oxygen exchange may also produce similar results.

Summary
Steady-State Transient Kinetic Analysis (SSITKA) of CO oxidation and CO-NO reaction over a commercial three-way auto catalyst indicated that CO adsorption/desorption was very fast or that the desorption step was negligible. The distribution of active sites for the formation of CO_2 was found to be bimodal for both reactions. For CO oxidation and CO-NO reaction less than 30% of the active sites accounted for more than 75% of the catalyst activity. This bimodal distribution of the site activities could not be attributed only to the bimetallic nature of the three-way auto catalyst since a similar bimodal activity distribution was also observed when these reactions were carried out on a mono-metallic Rh/La_2O_3. Although the presence of two types of active metal sites on both catalysts could not be ruled out, the low activity "sites" (with longer lifetimes) could be due to reaction with the rare earth oxides on the support or to multiple adsorption-desorption processes on metal sites.

Acknowledgments
Funding for this work from General Motors Co. (AC Rochester Division), which also provided us with the three-way auto catalyst, is gratefully acknowledged.

Literature Cited.

1. Happel, J. *Isotopic Assessment of Heterogeneous Catalysis*, Academic Press, Inc., New York, 1986.
2. Biloen, P., and Sachtler, W.H.M., *Adv. Catal.*, 1981, *30*, 165.
3. Biloen, P., *J. Mol. Catal.*, 1983, *21*, 17.
4. Yang, C.-H., Soong, Y., and Biloen, P., *J. Catal.*, 1985, *94*, 306.
5. Zhang, X., and Biloen, P., *J. catal.*, 1986, *98*,468.
6. De Pontes, M., Yokomizo, G.H., and Bell, A.T., *J. Catal.*, 1987, *104*, 147.
7. Iyagba, E.T., Nwalor, J.U., Hoost, T.E., and Goodwin, J.G., Jr., *J. Catal.*, 1990, *123*, 1.
8. Nwalor, J.U., and Goodwin, J.G., Jr., *J. Catal.*, 1989, *117*, 121.
9. Peil, K., Goodwin, J.G., Jr., and Marcelin, G., *J. Phys. Chem.*, 1989, *93*, 5977.
10. Peil, K., Goodwin, J.G., Jr., and Marcelin, G., *J. Amer. Chem. Soc.*, 1990, *112*, 6129.
11. Oh, Se H., Fisher, G.B., Carpenter, J.E., and Goodman, D.W., *J. Catal.*, 1986, *100*, 360.
12. Cant, N.W., *J. Catal.*, 1980, *62*, 173.
13. Gallaher, G.R., Goodwin, J.G., Jr., and Guczi, L., *Appl. Catal.*, 1991, *73*, 1.
14. Hecker, W.C., and Bell, A.T., *J. Catal.*, 1983, *84*, 200.
15. Cho, B.K., and Stock, C.J., *J. Catal.*, 1989, *117*, 202.

RECEIVED January 7, 1992

Chapter 6

Particle Size and Support Effects in NO Adsorption on Model Pt and Rh Catalysts

G. Zafiris, S. I. Roberts, and R. J. Gorte

Department of Chemical Engineering, University of Pennsylvania, Philadelphia, PA 19104

The adsorption of NO on Rh particles supported on $ZrO_2(100)$ and α-$Al_2O_3(0001)$ and on Pt particles supported on CeO_2, α-$Al_2O_3(0001)$, and the Zn- and O-polar surfaces of $ZnO(0001)$ has been studied using temperature programmed desorption (TPD). Metal particles were formed by vapor deposition, and the film morphology and particle formation were studied by Auger electron spectroscopy (AES) and transmission electron microscopy (TEM). For Rh, the support affected the stability of the metal films and the final particle size but made no observeable changes in NO adsorption properties. However, the TPD curves were found to be strongly dependent on the particle size for Rh. A low-temperature N_2 feature which is observed on bulk metals and large particles was not seen with small Rh particles. For Pt, both support composition and particle size influenced the TPD results. The results for CeO_2 and α-$Al_2O_3(0001)$ were essentially identical, and the changes in the NO TPD curves appear to be due to particle size only. An upward shift from 460 to ~550K was observed in the N_2 peak temperature when the particle size decreased below ~2.5 nm. Changes in the TPD curves for NO from Pt on the ZnO surfaces were more dramatic. Significant amounts of N_2O were observed in TPD for small particles and a N_2 desorption peak was observed at ~650K. The implications of these results on particle size and support effects in NO reduction catalysis will be discussed.

The main active components in automotive, emissions-control catalysts are noble metals supported on porous oxides. While attempts have been made to use less expensive materials, these attempts have been largely unsuccessful. However, it has been demonstrated that catalytic properties can be altered by changing the support structure and composition. Exactly how contact with an oxide can modify the

0097–6156/92/0495–0073$06.00/0

properties of the noble metals is not well understood; therefore, work in our laboratory has been directed toward gaining a better understanding of the nature of support-metal interactions so that support materials can be engineered to obtain optimal catalytic properties.

Our approach to understanding metal-support interactions has been to use model catalysts in which small metal particles are deposited onto oxide substrates. These model catalysts have several important advantages over typical, high-surface-area catalysts. First, surface composition and structure can be monitored much more easily using techniques like Auger electron spectroscopy (AES) and transmission electron microscopy (TEM). Second, temperature programmed desorption (TPD) can be easily interpreted since diffusion and readsorption will not influence the results.(1,2)   This ability to measure adsorption, desorption, and surface decomposition kinetics with TPD is of special interest with NO reduction catalysts. It has recently been demonstrated that macroscopic reaction rates for NO reduction by CO on Rh(111) can be accurately modeled using the microscopic rate parameters measured in TPD experiments.(3) Furthermore, large variations in the specific rates of NO reduction with particle size on alumina-supported Rh(4) were correctly predicted by TPD measurements of NO from model Rh/Al_2O_3 catalysts.(5)

In this paper, we will review previous work from our laboratory which characterized the structure of Rh films on alumina(5,6) and zirconia(7) and Pt films on alumina,(8) ceria,(9) and zinc oxide.(10)   This previous work demonstrated that relatively strong bonds can be formed between Pt or Rh and some oxides and that these interactions can affect the structure of metal particles and their adsorption properties.   We will then discuss recent TPD results for NO from each of these surfaces. The TPD curves indicate that support composition can significantly alter the desorption temperatures and products for small metal particles. The effect that the altered desorption kinetics could have on catalyst properties will be discussed.

Experimental

All TPD measurements were performed in a UHV chamber with a cylindrical mirror analyzer for AES, a metal source for evaporating Pt or Rh onto the oxide crystals, a calibrated film-thickness monitor for measuring metal coverage, and a quadrupole mass spectrometer mounted inside a stainless-steel cone to allow directed desorption from the sample. The oxide crystals were mounted on a Ta foil which could be resistively heated, and the temperature of the crystals was monitored using chromel-alumel thermocouples which were attached to the back side of the crystals using a ceramic adhesive. All TPD measurements were carried out with a heating rate of ~6K/sec and most experiments used labelled ^{15}NO to distinguish reaction products. Since the samples could not be heated to temperatures high enough to remove chemisorbed oxygen from Pt or Rh, the samples were heated to 770K in 5×10^{-8} Torr H_2 following each TPD experiment.

The TEM experiments were performed on a Philips EM 400T with an accelerating voltage of 120kV. The oxide crystals were first thinned by mechanical grinding and polishing, after which they were dimpled to a thickness of ~20μm. The crystals were then ion milled at 78K to perforation. After checking the samples in the microscope, they were placed in the same UHV chamber used in the TPD measurements, cleaned, and dosed with a known coverage of Pt or Rh. The samples were then carried in air to the microscope.

The α-$Al_2O_3(0001)$, $ZrO_2(100)$, and $ZnO(0001)$ samples were purchased as oriented single crystals. The Zn-polar surface of $ZnO(0001)$ was distinguished from

the O-polar surface by etching in concentrated HNO_3 as described elsewhere.*(11)*
The cerium oxide surface was prepared by spray pyrolysis onto an
α-Al_2O_3(0001) crystal held at 200°C, using an aqueous solution of $Ce(NO_3)_3$. The
ceria films showed only cerium, lanthanum, and oxygen in AES but were amorphous
and macroscopically rough, as shown by TEM characterization of a film deposited
onto a NaCl crystal which was then dissolved away.*(9)* While the ceria films were
continuous, TEM showed strong variations in the thickness of the film.
    The deposition of Pt and Rh onto each of the different oxides was carried out at
300K. Metal particle sizes were calculated from the metal coverage and the amount of
adsorbed CO, as determined by TPD. Particle sizes of selected samples were checked
using TEM and found to be in excellent agreement with those calculated from
adsorption, as discussed elsewhere.*(10)*

Results

Characterization of Pt and Rh Films. Since characterization of the metal films
on each of the various oxide substrates has been discussed in detail in other
publications,*(5-10)* we will only briefly describe the most important features. On α-
Al_2O_3(0001), three-dimensional particles of both Pt and Rh are formed following
deposition at room temperature.*(6)* This was demonstrated by following the AES
peak heights as a function of metal coverage and also by TEM. For Pt, transmission
electron diffraction (TED) showed that the metal particles formed by deposition were
randomly oriented. Furthermore, while the TPD curves for CO from Pt and Rh
particles were found to change with metal particle size, all of the observed desorption
features could be related to similar features found on Pt and Rh single crystals. These
results suggest that both Pt and Rh interact weakly with the α-Al_2O_3(0001) surface,
making this oxide substrate a good reference by which to compare results for the
other oxides.
    Evidence for relatively strong interactions was found for both Pt and Rh on
ZrO_2(100).*(7,18)* For both metals, AES indicated that film growth at 300K is close
to two-dimensional, although TEM results imply that irregularly shaped islands are
formed at this temperature, if not in vacuum, then upon exposure to air. Heating of
the metal films to approximately 500K caused the islands to agglomerate into larger
particles. A most interesting feature of the Pt and Rh particles is that they are
preferentially oriented with respect to the zirconia substrate. Well defined streaks are
observed in the TED pattern which demonstrate that the (111) surface of the metals
are in contact with the oxide and that the particles align themselves with the oxide.
The TPD results for CO from the Pt and Rh films were inconclusive regarding the
question of whether the interaction between the metals and zirconia results in changed
chemisorption properties. While TPD peak temperatures for CO from small Pt
particles on ZrO_2(100) are shifted downward compared to that observed for similarly
sized particles of Pt on α-Al_2O_3(0001), no peak shifts were observed for Rh
particles. Since desorption temperatures are strongly affected by crystallographic
orientation on Pt but are unaffected on Rh, it is uncertain whether the differences for
small Pt particles on ZrO_2 are due to chemical interactions between the oxide and the
metal or to changes in the orientation of the Pt particles.
    There is evidence for relatively strong interactions between Pt and ZnO.*(10)*
For room temperature deposition of Pt, two-dimensional films are formed which are

stable up to ~500K. TEM results again indicate that the Pt films, as well as the Pt particles which are formed by heating, are oriented with respect to the oxide substrate. Again, shifts are observed for CO desorption from Pt at low metal coverages, but these shifts appear to be due to chemical interactions between Pt and the oxide substrate. The CO peak shifts are found to be different for the O- and Zn-polar surfaces, suggesting that contact between the Pt atoms and the Zn^{+2} cations is responsible for affecting the adsorption properties.

The results for Pt on ceria are essentially identical to results for Pt on oxidized Al, which suggests relatively weak interactions.(9) While growth of the Pt film is two-dimensional at room temperature, this could be due to the fact that the ceria films prepared for our studies are macroscopically rough. We observed similar, two-dimensional film growth for Pt on an oxidized Al foil, but the TPD curves obtained for Pt on the oxidized foil were identical to those obtained for Pt on α-$Al_2O_3(0001)$.(8) Also, the Pt film coalesced into three-dimensional particles in exactly the same temperature range on both ceria and oxidized Al and gave similar particle sizes. Finally, the TPD curves for CO and H_2 from Pt particles of a given size are identical on both ceria and oxidized Al or α-$Al_2O_3(0001)$. While we did not examine the effect of pretreatment conditions on the ceria, our study indicates that the interactions between Pt and ceria are much weaker than those between Pt and either zirconia or zinc oxide and are similar to that between Pt and alumina.

NO Adsorption on Rh/ZrO$_2$(100) and Rh/α-Al$_2$O$_3$(0001). The results for NO adsorption on model Rh catalysts have been discussed in detail elsewhere(5) and will only be summarized here. What we observe are rather large changes in the TPD curves for NO as a function of particle size but relatively little change due to the substrate composition. This is shown in Figure 1, which gives the saturation TPD curves for NO from small and large annealed particles on α-$Al_2O_3(0001)$ and $ZrO_2(100)$. While the TPD features are better defined for Rh/ZrO$_2$ compared to those for Rh/Al$_2$O$_3$ due to improvements made in our experimental apparatus, the curves are essentially the same. For the largest particles (~10nm), the TPD curves are similar to that observed for Rh single crystals, with the exception that a larger fraction of the NO dissociates.(5) NO desorbs from a single feature at approximately 430K, while the N$_2$ desorbs from two well-defined features at 440K and between 450 and 700K. The low-temperature, N$_2$ peak has been identified as being due to the reaction NO_{ad} + N_{ad} = N_2 + O_{ad}. The second broad feature has been assigned to the recombination reaction, $2N_{ad} = N_2$, a process which occurs at different temperatures depending on the Rh crystallographic orientation.(12-15) For the smaller Rh particles (~3nm), all of the NO dissociates and only the nitrogen recombination peak is observed.

The fact that the TPD results for Rh on both $ZrO_2(100)$ and α-$Al_2O_3(0001)$ are so similar indicates that chemical modification of the Rh by the oxide substrate is probably not important. However, the large changes due to particle size probably are significant. It has been reported that NO reduction by CO is very structure sensitive on Rh.(4) Specific reaction rates were shown to increase by a factor of 45 in going from small to large Rh particles. Based on our TPD results, the increase in reaction rates on large Rh particles is likely due to the enhanced N$_2$ desorption rates indicated by the lower desorption temperatures. Auger spectra obtained from a Rh(111) crystal following high-pressure reaction measurements showed that the surface was covered with adsorbed nitrogen, which apparently prevents the adsorption of additional reactants.

It should be noted that, because NO reduction on Rh is structure sensitive, the oxide substrate may have an indirect effect on the catalyst. The TEM results discussed earlier demonstrate that zirconia can orient the Rh particles, which implies that the structure of the Rh particles will be changed. Therefore, for a given Rh particle size, the adsorption sites with the zirconia substrate may be different from the adsorption sites with an alumina substrate. Indeed, there is some evidence from the TPD results that this may be the case. It appears that the particle size at which the low-temperature, N_2 desorption feature at 440K is no longer present is >6 nm for Rh on $ZrO_2(100)$ and ~4 nm for Rh on α-$Al_2O_3(0001)$. While additional evidence is needed to prove that the support can influence structure-sensitive reactions by modification of the shape of the metal particles, this possibility should be considered when studying support effects in structure-sensitive reactions.

NO Adsorption on Pt/CeO$_2$ and Pt/α-Al$_2$O$_3$(0001).

Again, the results for Pt on ceria and α-$Al_2O_3(0001)$ have been discussed in detail elsewhere and will only be summarized.(9,16) Evidence from the film growth modes and from TPD of CO suggests that Pt interacts relatively weakly with both ceria and α-$Al_2O_3(0001)$. In agreement with this, the TPD results for NO from Pt particles supported on these two oxides are virtually identical. Figure 2 shows a comparison of TPD curves obtained for small and large Pt coverages on both substrates following a saturation exposure of NO. For a given metal coverage, the particles formed on ceria after annealing were larger than those formed on α-$Al_2O_3(0001)$. Therefore, in order to compare particles of similar size on the two substrates, the TPD curves from Pt/ceria are reported for the unannealed films, using the same metal coverages as that used to form the particles on the α-$Al_2O_3(0001)$. The results show that a small amount of NO desorbs unreacted near 440K for Pt on both substrates, along with a similar amount of N_2O in the Pt/ceria catalysts; however, in all cases, most of the NO dissociated during TPD to form N_2 and adsorbed oxygen. For large Pt coverages, the N_2 desorbs from a fairly sharp peak at 460K which is also observed for TPD of NO on Pt foils. For very small Pt coverages, a second N_2 desorption feature centered at 540K appears and the feature near 460K diminishes. It should be noted that the changes in the TPD curves occur at approximately the same coverage on both ceria and α-$Al_2O_3(0001)$, implying that there are no changes in the Pt adsorption properties due to the presence of ceria.

However, the changes in the NO adsorption properties with particle size are significant. From studies of NO coadsorbed with CO which have been reported earlier,(16) the desorption of N_2 on these Pt particles appears to be limited by the NO decomposition reaction. This implies that dissociation of NO is strongly affected by Pt particle size, which, in turn, may be very important in NO reduction catalysis. Rh is included in emissions-control catalysts because Pt is not as active for NO reduction by CO over the range of reactant pressures used in the catalytic converter. However, kinetic studies on a polycrystalline, Pt wire have reported rates that are comparable to those observed on Rh.(17) Our results suggest that small particles of Pt may not be as active as bulk Pt samples.

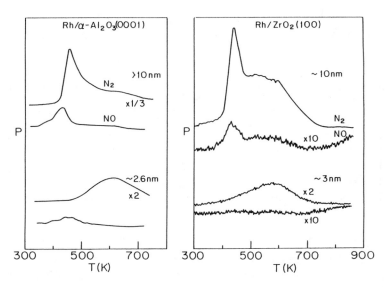

Fig 1)   TPD curves for NO from small and large particles of Rh on $ZrO_2(100)$ and α-$Al_2O_3(0001)$.

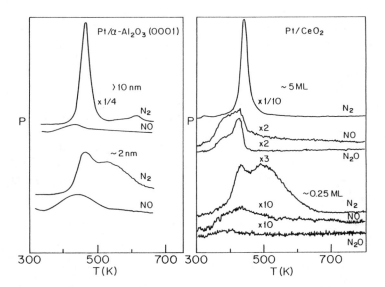

Fig 2)   TPD curves for NO from small (0.25ML) and large (5.0ML) coverages of Pt on CeO_2 and α-$Al_2O_3(0001)$. In order to form similar particle sizes on both substrates, the Pt films were not annealed prior to adsorption of NO.

It is also interesting to note that we observed a small amount of N_2O desorbing from the Pt/ceria sample. It is likely that a similar amount of N_2O was formed on the α-Al_2O_3(0001) sample, but that we have observed it in our current measurements due to improvements in our experimental capabilities. We hold off additional discussion N_2O for the next section.

NO Adsorption on Pt/ZnO(0001)O and Pt/ZnO(0001)Zn. The clearest evidence for strong support effects that we have observed in our studies has been with the two ZnO surfaces. In agreement with this, we also observe significant differences in the NO TPD curves for small Pt particles. Unlike our observations with CO, TPD curves for NO on the Zn- and O-polar surfaces were identical. All of the figures shown were obtained on the Zn-polar surface but results on the O-polar surface were the same.

Figure 3a) shows the TPD curves following saturation exposures of ^{15}NO on high ($5\times10^{15}Pt/cm^2$ or ~ 5 monolayers) and low ($0.5\times10^{15}Pt/cm^2$) coverages of Pt. The Pt films were not annealed prior to obtaining these results so that the films were initially two-dimensional prior to adsorption of NO and subsequent heating. For the high Pt coverage, the curves are virtually identical to the results discussed with ceria and alumina. Most of the NO dissociates to form N_2 which desorbs at 460K. Since the surface Pt is not in contact with the oxide substrate at these larger Pt coverages, the ZnO appears to have no influence on the adsorption results. For the lower Pt coverages, the results are significantly different. First, substantial fractions of the NO desorbed as N_2O (~ 15%) and unreacted NO (~ 7%), a result which contrasts sharply with results for high Pt coverages where N_2O and NO desorption accounted for ~1% and ~2% of the NO respectively. Second, the remaining N_2 now desorbs in two broad features at 450 and 650K. This second desorption peak is considerably higher than any feature observed with Pt on either ceria or alumina, which implies that it must be due to contact between Pt and ZnO. We attempted to determine whether the desorption of N_2 was limited by NO decomposition or by the recombination of adsorbed nitrogen by directing a beam of H_2 at the surface to monitor the point at which water was formed, but the results were inconclusive.

It was also not possible to determine the origin of the two N_2 desorption features in TPD. We examined lower Pt coverages to see if the feature at higher temperatures would become a larger fraction of the desorbing N_2 at still lower Pt coverages; however, the area under the two TPD features remained in approximately the same ratio for all Pt coverages in the submonolayer regime. Also, because the Pt film formed particles during the TPD measurement, it was not simple to vary the NO exposure to determine whether the two features filled sequentially. After each TPD experiment, the Pt had to be removed and a fresh layer deposited onto the oxide. Therefore, the two desorption states could be due to either two different types of sites or to repulsive interactions between adsorbed NO molecules. Whichever of these two mechanisms is responsible for the presence of the two desorption states, the fact that large amounts of N_2O form below 500K indicate that the surface processes responsible for the feature at lower temperatures are probably not the same as those which occur for Pt particles on alumina or ceria. Finally, it should be remembered that the Pt film begins to form particles at approximately 500K as the N_2 desorbs. It is not clear what role the movement of Pt atoms has on the desorption process,

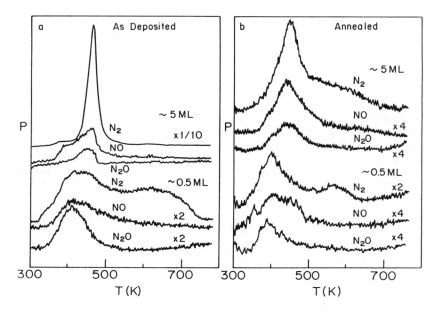

Fig 3) a) TPD curves for NO from Pt on ZnO(0001)Zn prior to annealing the Pt to form particles. b) TPD curves for NO from annealed Pt particles on ZnO(0001)Zn

although agglomeration of the Pt film also occurred with $5 \times 10^{15} Pt/cm^2$ during the TPD run without causing a high-temperature, N_2-desorption feature.

It is of interest to consider the nature of the sites on Pt/ZnO which lead to N_2O formation and the high-temperature N_2 feature. They are associated with the Pt atoms which are in contact with the ZnO but are unaffected by whether that Pt is in contact with the Zn- or O-polar surface of ZnO(0001). In a previous paper, we have argued that contact with the Zn^{+2} cations was critical based on differences in the TPD curves obtained for CO from Pt films on the two polar surfaces.(10) This does not appear to be a factor in NO adsorption. It is possible that the sites of interest are present at the edges of Pt islands on the ZnO and even that the adsorbed nitrogen and oxygen formed after NO dissociation may partially spillover onto the oxide. This is obviously very speculative and additional characterization is needed to determine the surface processes which occur on the low-coverage, Pt films.

The TPD curves obtained from the Pt films after annealing are shown in Figure 3b). Annealing causes the Pt film to coalesce into large particles, which in turn causes a significant decrease in the adsorption capacity. The main feature in the TPD curves for NO is the N_2-desorption state near 460K, which is essentially identical to that observed for the TPD curves from Pt on either alumina or ceria. The size of the particles formed from a given Pt film coverage is larger on ZnO than on alumina or ceria;(10) therefore, the large-particle desorption feature dominates even for low Pt coverages. This is further evidence that contact between the Pt and the oxide phase is necessary for the features observed for NO adsorption on the low-coverage, unannealed Pt films. Long-range, electronic considerations do not appear to have any effect on the adsorption properties of Pt. However, the presence of a high-temperature shoulder near 550K on the N_2 peak, observed for both large and small Pt coverages, does appear to be unique to the Pt particles supported on ZnO. This may be the result of NO adsorbed at interface sites between the Pt particles and the substrate. If this is the case, it is very interesting since it suggests that a substantial fraction of the adsorption sites may be modified by the support, even for large metal particles.

Discussion

The results for NO on model Pt and Rh catalysts have interesting implications for emissions-control catalysts. Significant changes in the decomposition of NO are observed as a function of both particle size and support composition. Unfortunately, none of the changes that we have observed point toward improved catalytic performance. Based on our previous results,(5) we suggest that NO reduction on Rh is limited by N_2 desorption. However, increases in Rh dispersion result in lower N_2 desorption rates and have been shown to lead to significant decreases in the specific rates for NO reduction by CO.(4) For Pt, it appears that NO reduction is limited by NO dissociation. Again, increases in Pt dispersion lead to decreases in the expected catalytic performance since the NO decomposition reaction occurs at higher temperatures on small particles. Also, the supports we investigated had either no effect on NO adsorption or decreased the rates of the desired steps. The demonstration, however, that the support composition can affect the adsorption properties does hold the promise that some oxides may enhance catalytic properties.

The results from model catalysts may give clues as to which types of oxides hold promise. In previous papers, our group has argued that support effects are due to contact between the catalytic metal and the support cations based on results from TPD of CO.(10,18) The results for NO adsorption on Pt provide partial support for this picture. Support effects were observed for ZnO but not for ceria or alumina. It is likely that the Ce^{+4} and Al^{+3} cations in ceria and alumina are surrounded by oxygen so that the Pt does not have direct contact with the oxide cations. With ZnO, the cations are much more available. However, it is not clear why we did not observe differences between ZnO(0001)Zn and ZnO(0001)O surfaces based on this model. Additional characterization of this adsorption system could provide important clues as to how one might manipulate the catalyst support in order to enhance catalytic performance for NO reduction.

Acknowledgements

This work was supported by the DOE, Basic Energy Sciences, Grant #DE-FG03-85-13350. Support for the electron microscopy experiments was partially provided by the NSF, MRL Program, Grant #DMR 88-19885. We would also like to thank Dave Luzzi for his technical assistance with the electron microscopy.

Summary

Particle size and support composition can play an important role in the decomposition of NO and subsequent desorption of N_2 on Pt and Rh catalysts. For Rh, NO reduction appears to be structure sensitive due to changes in the N_2 desorption rates. Support and particle size appear to influence the structure of the adsorption sites, which in turn affects desorption rates. On Pt, NO decomposition appears to limit reaction rates. The decomposition appears to be affected by both the structure of the sites and the proximity of oxide cations.

Literature Cited

1)   Gorte, R.J., *J. Catal.* **1982**, *75*, 164.
2)   Demmin, R.A., and Gorte, R.J., *J. Catal.* **1984**, *90*, 32.
3)   Oh, S.H., Fisher, G.B., Carpenter, J.E., and Goodman, D.W., *J. Catal.* **1986**, *100*, 360.
4)   Oh, S.H., and Eickel, C.C., *J. Catal.* **1991**, *128*, 526.
5)   Altman, E.I., and Gorte, R.J., *J. Catal.* **1988**, *113*, 185.
6)   Altman, E.I., and Gorte, R.J., *Surf. Sci.* **1988**, *195*, 392.
7)   Zafiris, G., and Gorte, R.J., *J. Catal.*, **1991**, *132*, 000.
8)   E. I. Altman and R. J. Gorte, *J. Catal.* **1988**, *110*, 191.
9)   Zafiris, G., and Gorte, R.J., to be published.
10)   Roberts, S., and Gorte, R.J., *J. Chem. Phys.* **1990**, *93*, 5337.
11)   Mariano, A.N., and Hanneman, R.E., *J. Appl. Phys.* **1963**, *34*, 384.
12)   Root, T.W., Schmidt, L.D., and Fisher, G.B., *Surf. Sci.* **1985**, *150*, 173.
13)   Baird, R.J., Ku, R.C., and Wynblatt, P, *Surf. Sci.* **1980**, 97, 346.
14)   Ho, P. and White, J.M., *Surf. Sci.* **1984**, *137*, 103.
15)   Thiel, P.A., Williams, E.D., Yates, J.T., and Weinberg, W.H., *Surf. Sci.* **1979**, *84*, 54.
16)   Altman, E.I., and Gorte, R.J., *J. Phys. Chem.* **1989**, *93*, 1993.
17)   Klein, R.L.; Schwartz, S.; Schmidt, L.D., *J. Phys. Chem.* **1985**, *89*, 4914.
18)   Roberts, S.A., and Gorte, R.J., *J. Phys. Chem.* **1991**, *95*, 5600.

RECEIVED December 19, 1991

Chapter 7

Effect of Ce on Performance and Physicochemical Properties of Pt-Containing Automotive Emission Control Catalysts

J. G. Nunan[1], Ronald G. Silver[2], and S. A. Bradley[3]

[1]Allied-Signal Research and Technology Center, P.O. Box 5016, Des Plaines, IL 60017–5016
[2]Allied-Signal Automotive Catalyst Company, P.O. Box 580970, Tulsa, OK 74158–0970
[3]UOP Research Center, P.O. Box 5016, Des Plaines, IL 60017–5016

Present-day automotive emission control catalysts contain noble metals such as Pt, Pd and Rh all on an alumina support with a variety of promoters. Ce is one of the most important promoters. The interaction between Pt and Ce was studied using TPR and STEM on a variety of catalysts. The degree of Pt/Ce interaction was increased by decreasing CeO_2 crystallite size, and to a lesser extent by increasing CeO_2 loading. Direct Pt/Ce interaction leads to a synergistic reduction of both Pt and surface Ce. This reduction qualitatively correlates with catalyst performance after activation in a reducing gas. It is proposed that this synergistic reduction of Pt and Ce is associated with observed improvements in catalyst performance using a non-oscillating exhaust gas.

Current state-of-the-art automotive emission control catalysts simultaneously oxidize hydrocarbons (HCs) and carbon monoxide (CO), and reduce oxides of nitrogen, when operated under conditions near the stoichiometric point. The catalysts generally consist of noble metals (NMs) such as Pt, Pd, and Rh, on alumina supports. The addition of CeO_2 to these catalysts promotes their activity, and has been well studied (1-7). Among its benefits, the added CeO_2 may serve as an "oxygen storage" component under the cyclic lean-rich composition fluctuation common to automotive engine exhaust gas held near the stoichiometric point. The CeO_2 component provides oxygen, which is stored under lean conditions, for the oxidation of HC and CO under rich conditions (1, 8-11). CeO_2 is also thought to provide other functions, such as stabilization of the alumina (12), promotion of the water-gas-shift reaction under rich conditions (2), and promotion of noble metal dispersion (3,6,13,14). From a reaction standpoint, small amounts of CeO_2 have been proposed to promote CO oxidation activity (15,16), which may lead to higher performance for HC and NOx conversion.

0097–6156/92/0495–0083$06.00/0

Another aspect of CeO_2 addition to automotive catalysts is its effect on the oxidation states of the NMs. CeO_2 promotes the reduction of the NMs with which it comes in contact (1,12) under reducing conditions. In addition, Yao and Yao(1) have observed the synergistic promotion of the reduction of CeO_2 in the presence of Pt or Rh.

The benefits of CeO_2 have not only been observed for fresh catalysts, but also for catalysts aged under severe conditions on an engine dynamometer for as long as 100 hours (17). The durability of these CeO_2 promotion effects emphasizes the importance of further study of the role CeO_2 plays in improving automotive exhaust emission control.

Previous work in this area has tended to focus either on model studies where a limited number of reactants were present during catalyst testing, or on studies where the CeO_2 containing catalysts were evaluated using dynamometer aging and testing conditions. In a recent model study in our laboratories (18), using a full complement exhaust gas mixture, it was shown that direct Pt/Ce interaction led to large beneficial effects on catalyst performance after activation of the catalyst in rich or stoichiometric exhaust gas mixtures. The primary impact of CeO_2 on fresh and laboratory aged catalysts was on the CO oxidation activity, which led to improved HC and NOx conversion. The improved activity of the CeO_2 containing catalysts qualitatively correlated with the extent of the synergistic reduction of Pt and surface Ce, as probed by temperature programmed reduction (TPR). The present study attempts to further explore the relationship between the effect of CeO_2 on the activity of Pt containing emission control catalysts, and the synergistic Pt-Ce reduction feature observed over these catalysts.

Experimental.

The compositions of the catalysts used in this study are listed in Table I.

TABLE I. Catalyst Composition and CeO_2 Crystallite Size

Catalyst Description	CeO_2 Size(Å)	Content (wt%)	
Pt/γ-Al$_2$O$_3$	—	0.86 Pt	
Pt,24%Ce/γ-Al$_2$O$_3$	~65	0.77 Pt	
Pt,Rh/γ-Al$_2$O$_3$	—	0.745 Pt,	0.045 Rh
Pt,Rh,5%Ce/γ-Al$_2$O$_3$	67	0.77 Pt,	0.04 Rh
Pt,Rh,6%Ce/γ-Al$_2$O$_3$	213	0.79 Pt,	0.041 Rh
Pt,Rh,6%Ce/γ-Al$_2$O$_3$	332	0.78 Pt,	0.04 Rh
Pt,Rh,6%Ce/γ-Al$_2$O$_3$	1000	0.82 Pt,	0.043 Rh
Pt,Rh,24%Ce/γ-Al$_2$O$_3$	~65	1.05 Pt,	0.038 Rh
Pt,Rh,25%Ce/γ-Al$_2$O$_3$	81	0.63 Pt,	0.035 Rh
Pt,Rh,25%Ce/γ-Al$_2$O$_3$	1000	0.82 Pt,	0.043 Rh

Granular Catalyst Studies. This study utilized model granular catalysts that contained Pt. Some of these catalysts also contained Rh and most contained varying amounts of CeO_2. In those catalysts containing Pt and CeO_2, the degree of Pt/Ce interaction was varied by controlling both CeO_2 loading and crystallite size. Samples containing highly dispersed CeO_2 with extensive Pt/Ce interaction were prepared by impregnation of γ-Al_2O_3 with Ce salts, followed by calcination in air at 600°C for 6 hours. Samples with low CeO_2 dispersion and having poor Pt/Ce interaction were prepared by combining the γ-Al_2O_3 support with CeO_2 powders that had been sintered to varying degrees. Pt and Rh were introduced by impregnation with aqueous solutions of H_2PtCl_6 and $RhCl_3$. The Pt and Rh loadings for all granular catalysts are shown in Table I.

For activity testing, one gram of catalyst was evaluated in a full complement exhaust gas at a total feed rate of 5 l/min. A rich and a stoichiometric exhaust gas mixture were each used during the course of catalyst evaluations. The rich mixture consisted of 8000 ppm CO, 2667 ppm H_2, 267 ppm C_3H_8, 2790 ppm O_2, 1835 ppm NO, 167 ppm C_3H_6, 11.88% CO_2, and 10% water with the balance N_2. The stoichiometric mixture consisted of 5775 ppm CO, 1925 ppm H_2, 193 ppm C_3H_8, 4645 ppm O_2, 1835 ppm NO, 167 ppm C_3H_6, 11.88% CO_2, and 10% water with the balance N_2. The testing profile consisted of heating the catalyst in the stoichiometric exhaust at 5°C per minute to 450°C (Rise-1), holding at 450°C in the stoichiometric or rich gas for 30 minutes, followed by a drop to 50°C. This was followed by a second rise to 450°C (Rise-2) in the stoichiometric exhaust gas mixture.

Monolith Catalyst Studies. This study utilized monolith catalysts prepared using a commercial process. In this case, the Pt/Ce interaction was varied by changing the CeO_2 loading. Catalyst washcoats were prepared containing varying relative amounts of CeO_2 and γ-Al_2O_3, and applied to cordierite monoliths having 64 square cells/cm^2. The nominal Pt loading for these catalysts, added using the chloride salt, was 0.6 wt%. Catalyst on monolith supports, such as these, are more typical of an actual vehicle catalyst than are the granular catalysts described above.

For activity testing, 1 inch diameter cylindrical cores were removed from the monolith catalyst at several locations. Each core was sliced into a number of 0.5 inch long sections, then four sections were selected and assembled into a 2 inch long composite core. This was intended to limit problems due to possible uneven washcoat loading on the cordierite support. A similar feedgas to that used for granular catalyst evaluations was employed and 20 ppm SO_2 was also added to the mixture. Further, non-oscillating exhaust gases were used, as with the granular catalysts, in order to eliminate oxygen storage effects. The total gas flow rate was 20 l/min. The test procedures were also similar, except that steady state performance was measured at 315°C in one experiment and 275°C in the other.

In both the monolith and granular catalyst studies, samples were characterized using temperature programmed reduction (TPR), x-ray diffraction (XRD), and scanning transmission electron microscopy (STEM). For the TPR studies all catalysts were initially pretreated in 20%O_2/80%He at 600°C for 1 hour before the experiment. The electron microscopy was performed with a Vacuum Generators HB-5 dedicated STEM.

Results and Discussion.

Granular Catalyst Characterization. Catalysts containing varying loadings and dispersions of CeO_2 were characterized using TPR and STEM analysis, both before and after evaluation in the synthetic exhaust gas at 450°C. Table II summarizes the

TABLE II. Effect of CeO_2 Loading and Crystallite Size on the Degree of Pt-Ce Interaction over Pt,Rh,Ce/γ-Al_2O_3 Granular Catalysts

			Pt/Ce Ratios			
Ce Loading	Catalyst Tested?	CeO_2 Size(Å)	Nominal	(A) Calculated on CeO_2	(B) Observed on CeO_2*	Ratio (B)/(A)
5%	No	67	0.110	0.004	0.040	10
5%	Yes	67	0.110	0.004	0.060	15
6%	No	213	0.090	0.0013	0.014	11
6%	No	332	0.090	0.0008	0.008	10
25%	No	1000	0.022	0.0004	0.001	2.5
25%	No	81	0.022	0.0035	0.029	8.3
25%	Yes	81	0.022	0.0035	0.023	6.6

* Typical standard deviation is +/- 30%.

STEM analyses for a series of Pt/Rh catalysts having varying CeO_2 crystallite sizes and loadings. The CeO_2 loadings were varied between 5 and 25wt.%, and the crystallite sizes were varied between 60 and 1000Å. The nominal Pt/Ce ratio reported is the ratio between the total amount of Pt and the total amount of Ce added to the catalyst. The calculated Pt/Ce on CeO_2 is an approximate ratio based on the relative surface areas of the γ-Al_2O_3 and CeO_2 components of the catalyst and assumes even dispersion of the Pt over the total catalyst surface. The observed Pt/Ce ratio on CeO_2 was obtained from chemical analysis in the STEM, and is substantially higher than expected based on a random distribution of Pt throughout the sample, especially for samples containing moderate to small CeO_2 crystallite sizes. For the sample with the small CeO_2 crystallites and large CeO_2 loading, the total and observed Pt/Ce ratios are similar, indicating that most of the Pt is associated with the CeO_2. This is true even though CeO_2 accounts for only a small fraction of the total surface area of the catalyst. This association of Pt with the CeO_2 component remains even after catalyst testing. Pt crystallite sizes were also measured in the STEM for two catalysts after evaluation in the synthetic exhaust gas mixture. The tested catalyst with 5% Ce in the support had Pt crystallites ranging in size from 30 to 50Å, while the tested catalyst with 25% Ce in the support had Pt crystallites from 100 to 250Å. Also shown in the last column of Table II is the ratio of the observed to calculated ratios of Pt/Ce. It is seen that the ratios are much greater than 1.0 again indicating the selective association of Pt with the CeO_2 component.

Pt and Pt/Rh granular catalysts were characterized using TPR and the results are shown in Figures 1 to 3. The effect of both CeO_2 addition and the extent of NM/Ce interaction on the TPR spectrum is shown in Figure 1. The Pt/γ-Al_2O_3 sample shows a single broad peak centered at 260°C, associated with Pt reduction. The 25%Ce/γ-Al_2O_3 sample shows a small but continuous H_2 uptake feature above 100°C with a distinct maximum in uptake at 515°C. These features are associated with surface and possibly subsurface Ce reduction (*1,18*). Further H_2 uptake at higher temperatures is assigned to bulk Ce reduction. The sample containing both Pt and CeO_2 shows features completely different than the previously described catalysts. A single sharp reduction peak is observed at 175°C. The H_2 uptake associated with this peak is much larger than could be accounted for if it were just due to Pt reduction. No peaks for either Pt reduction on γ-Al_2O_3, or surface Ce reduction are observed. Thus, the peak at 175°C is assigned to the synergistic reduction of Pt and surface Ce (1,18) which occurs at lower temperatures when the two components are present together. For the Pt,Rh,25%Ce/γ-Al_2O_3 catalyst, a sharp reduction peak is again observed but now the peak is centered at 160°C. The peak intensity is also greater than that observed for the Rh-free sample indicating that Rh may further promote the reduction of Ce and Pt.

Figure 2 shows the effect of varying the extent of Pt/Ce interaction on the TPR spectra. This was done by varying both the CeO_2 crystallite size and loading as shown by the STEM analysis in Table II. Increasing the degree of Pt/Ce interaction, as shown by the ratio of observed to calculated Pt/Ce on CeO_2 in Table II, leads to a systematic increase in the intensity of the low temperature peak associated with synergistic reduction of NM and surface Ce. This increase comes about at the expense of the Pt on γ-Al_2O_3 reduction peak, and also the surface Ce reduction peak. It is also evident that the TPR spectra of both the freshly calcined and tested catalysts are very similar, indicating that exposure to the laboratory feedgas and consequent partial reduction and sintering of the Pt does not greatly affect the TPR spectra. The intensity of the low temperature reduction peak is, however, lower for the tested sample than for the fresh sample. This is probably due to sintering of the Pt, leading to partial segregation from the CeO_2 component. Partial sintering of Pt also seems to lead to more facile reduction as shown by the shift of the peak maximum to lower temperatures for the tested catalyst.

Figure 3 shows the TPR spectra for the two samples whose Pt crystallite size distributions were determined by STEM after activity testing. The CeO_2 crystallite size was in the range of 60 to 80Å for both samples. The sample with the higher Ce loading gave reduction at lower temperatures and a more intense synergistic reduction peak even though it had the larger Pt crystallite sizes after testing.

Granular Catalyst Activity Testing. Some of the catalysts characterized using TPR and STEM were also tested for HC and CO oxidation activity. Figure 4 shows a plot of CO conversion versus temperature (lightoff performance), for several catalyst formulations. The Ce-containing catalysts were activated in the rich feedgas during the hold between Rise-1 and Rise-2, or by pretreatment in a 5%H_2/95%N_2 stream at 450°C for 0.5 hours. In Figure 4 a large benefit in performance is observed for these catalysts after the rich exhaust gas or H_2 pretreatments. The benefit of Ce addition to the catalyst is only seen after activation has taken place, during the Rise-2 lightoff.

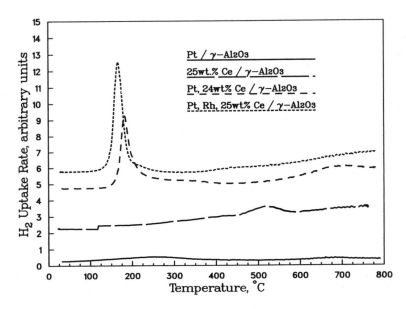

Figure 1.   Pt-Rh-Ce and Pt-Ce interaction results in synergistic reduction compared to individual components.

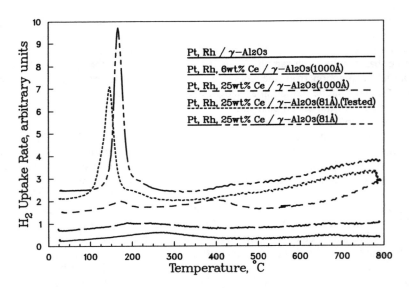

Figure 2. Increasing Pt-Ce synergistic reduction as a function of CeO₂ loading and crystallite size.

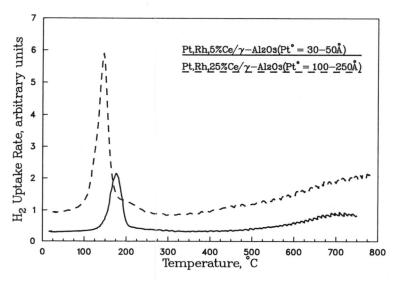

Figure 3. TPR spectra after testing for different Ce loadings (CeO$_2$ size = 60 - 80Å).

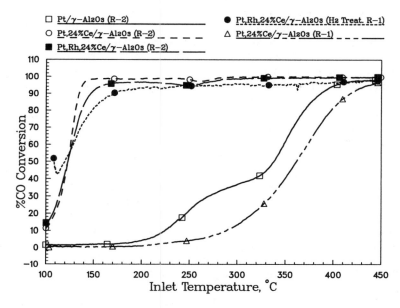

Figure 4. Effect of pretreatment and Pt-Ce interaction on catalyst lightoff performance.

Thus the Pt,24%Ce/γ-Al$_2$O$_3$ sample shows low activity for Rise-1, but then has dramatically improved performance during Rise-2, with the lightoff temperature decreasing by about 150°C. Similar large activations have been observed in the past as a result of rich or stoichiometric exhaust gas pretreatments and have been associated with reduction of both Pt and surface Ce (18). In contrast, the Ce-free Pt-only sample shows poor lightoff performance even for Rise-2 indicating that Ce is playing a key role in the performance of Ce containing catalysts. For the Pt,Rh,24%Ce/γ-Al$_2$O$_3$ catalyst the Rise-2 lightoff curve was found to be similar to the one observed for Pt,24%Ce/γ-Al$_2$O$_3$. This demonstrates that the performance and activation of these fresh catalysts is not unique to Pt-only catalysts and is not impacted to a large extent by the presence of Rh. The Rise-1 lightoff activity of the Pt,Rh,24%Ce/γ-Al$_2$O$_3$ catalyst that was prereduced in H$_2$, showed activity similar to the Rise-2 lightoff of a Ce containing catalyst that had been exposed to a rich exhaust gas between Rise-1 and Rise-2. This further supports the premise that activation is associated with catalyst reduction.

Figure 5 shows the effect of CeO$_2$ loading and crystallite size on the temperature required to obtain 25% conversion of either CO or HC. The lower the temperature the more active the catalyst. Results shown are from the Rise-2 lightoff performance, after the catalysts have been activated in the rich feedgas. Moderate improvements in HC and CO conversions are observed with increasing CeO$_2$ loading for catalysts with relatively large CeO$_2$ crystallites (i.e. poor Pt/Ce interaction). However a dramatic improvement is seen with decreasing CeO$_2$ crystallite size from 1000 to 81Å. The largest improvement is observed for CO which is consistent with other studies in the literature (4,14). The observed increases in catalyst activity with increased CeO$_2$ loading and decreased CeO$_2$ crystallite size, qualitatively correlate with the appearance and subsequent increase in intensity of the synergistic reduction peak observed in the TPR spectra for these same samples as seen in Figure 2. The growth of the synergistic reduction peak, in turn, correlates with the extent of Pt/Ce interaction as shown in Table II. Since little or no effect of CeO$_2$ loading or crystallite size was observed for the fresh lightoff activity, the presence of CeO$_2$ clearly has its dominant impact on catalyst activation in the exhaust gas. This activation in turn correlates with the extent of Pt/Ce interaction and the synergistic reduction of Pt and surface Ce.

Figure 6 shows the Rise-1 and Rise-2 CO lightoff activities for the Pt,Rh,5%Ce/γ-Al$_2$O$_3$ and Pt,Rh,25%Ce/γ-Al$_2$O$_3$ samples for which the TPR spectra were reported in Figure 3 after testing. The Pt crystallite size ranges for both samples as measured by STEM after testing are also shown in Figure 6. The Rise-1 lightoff profiles show an advantage for the sample with the smaller Pt crystallites (i.e. better dispersion) and lower Ce loading. However extensive activation occurred for both catalysts between Rise-1 and Rise-2 and the results for the second, and subsequent, rises now show the sample with the higher Ce loading and larger Pt crystallites (Pt,Rh,25%Ce/γ-Al$_2$O$_3$) having the highest performance. This is surprising as this sample contains substantially larger Pt crystallite sizes as compared to the Pt,Rh,5%Ce/γ-Al$_2$O$_3$ sample. The result suggests that activated catalyst performance is related more to the extent of Pt/Ce interaction, as characterized by the intensity of the synergistic reduction peak, than merely to Pt crystallite size.

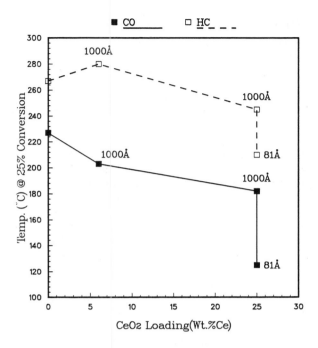

Figure 5. CeO₂ lightoff activity as a function of CeO₂ loading and crystallite size.

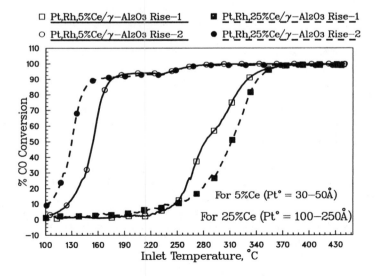

Figure 6. Catalyst performance as a function of loading (CeO₂ size = 60 - 80Å).

Monolith Catalyst Characterization. TPR spectra for a Ce-free and a 34wt.%Ce containing Pt catalyst are compared in Figure 7. Similar reduction features are observed for these samples as for the granular catalysts. The Pt/γ-Al$_2$O$_3$ sample displays a broad H$_2$ uptake peak centered at 300°C, the intensity of which correlates with Pt reduction, based on the analyzed Pt content. Addition of Ce clearly leads to an attenuation of the peak associated with Pt reduction, and to the appearance of H$_2$ uptake features at lower temperatures, as observed for the granular samples. There is no evidence of H$_2$ uptake features at higher temperatures associated with surface Ce reduction, as observed earlier for samples that do not contain NMs. This again suggests that the low temperature reduction feature is due to the synergistic reduction of Pt and surface Ce. However, the shift to lower reduction temperature and the intensity of the synergistic reduction feature is less than for the granular samples. The synergistic reduction feature is now observed at 200°C, and a distinct shoulder is seen at 250°C. This shoulder is probably due to reduction of Pt not associated with Ce. For these preparations, the Pt was probably not as selectively associated with the CeO$_2$ component. This is supported by STEM analyses which showed a lower degree of Pt/Ce association in these samples as compared to the granular catalysts.

Monolith Catalyst Activity Testing. The steady-state conversions of CO, HC and NO over the Pt/γ-Al$_2$O$_3$ and the Pt,34%Ce/γ-Al$_2$O$_3$ monolith catalysts are compared in Figure 8, after the catalysts were activated in a rich exhaust gas. Catalysts were tested in a manner similar to that used in testing the granular catalysts except that 20 ppm SO$_2$ was added to the feedgas and the temperature at the end of Rise-2 was reduced to 315°C. The steady state activity of the catalyst was then measured at this point. Both catalysts were tested using a non-oscillating feedgas, which eliminated the possibility of oxygen storage playing a role in the Ce-containing catalyst activity. Despite this lack of oxygen storage, conversions over the Ce-containing sample are 67% higher for CO, 56% higher for HC, and 60% higher for NO relative to the Ce-free sample.

Pt crystallite sizes were measured using STEM for both samples after testing. The Ce containing sample had smaller Pt crystallites, and also better conversions, as shown in Figure 8. Based on this observation it is not possible to unambiguously attribute the higher activity of the Ce containing catalyst to the synergistic reduction feature, as was the case for the lightoff performance results over granular catalysts.

In an effort to determine if the above improvements in performance with Ce addition could be explained by Ce promotion of the WGS reaction, the above experiment was modified by removing CO and CO$_2$ from the catalyst feedgas. The reaction temperature was also lowered to 270°C, as higher conversions could be achieved in the absence of the inhibiting effects of CO. A separate study over a Pt,Rh/CeO$_2$ catalyst had shown that the WGS reaction was effectively poisoned by SO$_2$ under these conditions. Figure 9 shows the steady-state HC and NO conversions. The Ce containing catalyst has the highest activity, suggesting that under these reaction conditions promotion of the WGS reaction cannot explain the observed benefits of Ce addition.

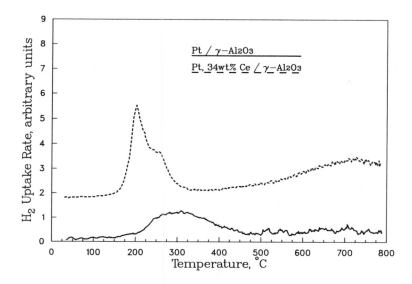

Figure 7. Pt-Ce synergistic reduction observed over monolith catalysts.

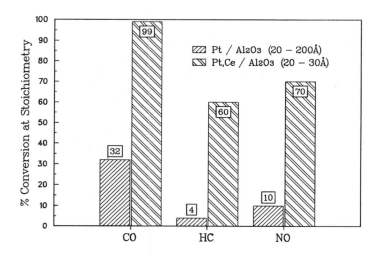

Figure 8. Steady state performance advantages of Pt-Ce over Pt.

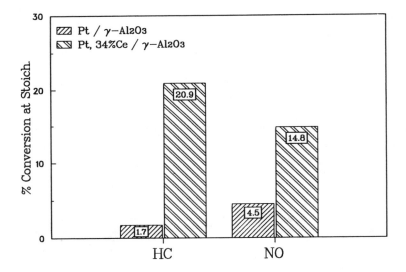

Figure 9. Steady state performance advantages of Pt-Ce over Pt in the absence of CO and CO_2. T = 270°C; Feedstream contains 20 ppm SO_2.

Conclusion.

There is a promotional effect of Ce addition to Pt containing automotive exhaust catalysts which extends beyond oxygen storage effects, water-gas-shift reaction promotion, and the promotion of Pt dispersion. This promotional effect is only observed after the catalyst has been activated by reduction in H_2 or a rich exhaust gas, and it appears to correlate with the synergistic reduction of Pt and surface Ce. The degree of Pt/Ce interaction can be increased by increasing the CeO_2 loading or decreasing the CeO_2 crystallite size. Increasing this interaction leads to an increase in the magnitude of the synergistic reduction and in the degree of activity promotion.

Acknowledgments.

The authors are indebted to their respective organizations for financial support, encouragement and provision of facilities. They acknowledge the following colleagues for their assistance at both Allied-Signal and UOP: P. J. Pettit and D. E. Mackowiak for catalyst preparation; M. T. West for testing the granular catalysts and D. H. Whisenhut for testing the monolith catalysts; M. J. Cohn for TPR analysis; M. A. Vanek and J. A. Triphahn for XRD analysis. They would also like to thank H. J. Robota, K. C. C. Kharas, W. B. Williamson, and M. G. Henk for helpful discussions

Literature Cited.

1. Yao, H. C. and Yao, Y. Y.-F., J. Catal. **1984,** 86, p. 254.
2. Kim, G., Ind. Eng. Chem. Prod. Res. Dev. **1982,** 21, p. 267.
3. Gandhi, H. S., and Shelef, M., Stud. Surf. Sci. Catal. **1987,** 30, p.199.
4. Yao, Y. Y. -F., J. Catal. **1984,** 87 , p. 152.
5. Yao, Y. Y. -F., Ind. Eng. Chem. Prod. Res. Dev. **1980,** 19, p. 293.
6. Yao, H. C., Appl. Surf. Sci. **1984,** 19, p. 398.
7. Shyu, J. Z. and Otto, K., J. Catal. **1989,** 115, p. 16.
8. Gandhi, H. S., Piken, A. G., Shelef, M., and Delosh, R. G., SAE Paper No. 760201 **1976.**
9. Su, E. C., Montreuil, C. N., and Rothchild, W. G., Appl. Catal. **1985,** 17, p. 75.
10. Rieck, J. S. and Bell, A. T., J. Catal. **1986,** 99, p. 278.
11. Hicks, R. F., Rigano, C. and Pang, B., Catalysis Letters **1990,** 6, p. 271.
12. Harrison, B., Diwell, A. F., and Hallett, C., Platinum Metals Rev. **1988,** 32, p. 73.
13. Sergey, F. J., Masellei, J. M., and Ernest, M. V., W. R. Grace Co., U. S. Patent 3903020 **1974.**
14. Hindin, S. G., Englehard Mineral and Chemical Co., U.S. Patent 3870455 **1973.**
15. Summers, J. C. and Ausen, S. A., J. Catal. **1979,** 58, p. 131.
16. Oh, S. H., and Eickel, C. C., J. Catal. **1988,** 112, p. 543.
17. Silver, R. G., Summers, J. C. and Williamson, W. B., in "Studies in Surface Science and Catalysis", (Catalysis and Automotive Pollution Control II), A. Crucq, eds., Elsevier, New York, **1991,** Vol. 71, pp. 167 - 180.
18. Nunan, J. G., Robota, H. J., Cohn, M. J., and Bradley, S. A., J. Catal. **1992,** 133, p. 309.

RECEIVED February 18, 1992

STATIONARY SOURCE EMISSION CONTROL

Chapter 8

The 1990 Clean Air Act and Catalytic Emission Control Technology for Stationary Sources

Jerry C. Summers, John E. Sawyer, and A. C. Frost

Allied-Signal Automotive Catalyst Company, P.O. Box 580970, Tulsa, OK 74158–0970

Chapter 1 of this book outlined the legislative requirements expected for controlling the emissions from light-duty vehicles over the next decade, and how these requirements are expected to be met, particularly through the anticipated advances in the development of catalysts and their supporting systems. Similarly, this chapter reviews the provisions of the 1990 Clean Air Act that deal with emissions from stationary sources, and the technologies that are available to meet these provisions, particularly catalytic incineration. The subsequent, technically specific chapters in this book are referenced throughout the text of this chapter as representations of the advances that are being made in the various areas of catalytic control of emissions from stationary sources.

Background Legislation. During the past twenty years air pollutants have been reduced by technology mandated by state and federal regulations. The federal regulations were based on four acts. The Clean Air Act of 1967 provided the first federal authority to establish air quality standards. The following Clean Air Act of 1970, while comprehensive for its time, was subsequently amended, first in 1977, and then again in 1990.

The initial primary approach for attacking the nation's air pollution problems was to regulate tailpipe emissions from light-duty vehicles. These efforts led to the automotive catalytic converter and unleaded gasoline. Other classes of vehicles were also brought under regulation. Additionally, the clean air acts established requirements for new sources of pollutants.

While these regulations have resulted in a 95 percent reduction of airborne lead, a 38 percent reduction in particulates and a 16 percent reduction in ozone, the overall air quality remains far short of the National Air Quality Standards(1). In fact, over 100 million Americans live in areas that exceed one or more of the standards(2).

Ozone, formed from a complex chemical reaction between volatile organic compounds (VOCs) and nitrogen oxides (NOx) in the presence of sunlight, is still the most pervasive pollutant. This is because the VOCs and

NOx levels have remained high. For example, while tailpipe VOC (commonly called hydrocarbon) and NOx emissions/mile have dropped significantly over the last twenty years, the total miles traveled has doubled(3). The number of stationary sources, whose yearly tonnage of VOC emissions closely match that of automobiles, has also increased. While the larger stationary sources were subjected to some control, the smaller ones (less than 100 tons/year) were not. Taken collectively, the smaller units emit unacceptable quantities of pollution in many urban areas.

While ozone aggravates chronic heart and respiratory diseases, another class of pollutants, air toxics, can cause serious long term problems. Air toxics are typically carcinogens, mutagens, and reproductive agents. Only seven of these toxics were regulated by the 1977 Clean Air Act. The amounts of the remaining, unregulated toxics are immense. Over 2.7 billion pounds are emitted annually in the United States, leading directly to an estimated 1000 to 3000 cancer deaths per year(2).

Less serious to human health, but still major environmental concerns, are the formation of acid rain and the reduction of the stratospheric ozone. Some of the approximately twenty-two million metric tons each of SO_2 and NOx emitted annually in the United States(4) are transformed in the atmosphere into their acids, which are carried to the earth with rain or snow; this acid rain damages poorly buffered lakes, harms forests and buildings, causes reduced visibility, and is suspected of damaging health. In another set of atmospheric reactions halogenated organic compounds react with the stratospheric ozone; the resulting depletion of this ozone layer is expected to increase skin cancer, cataracts and birth defects, as well as to reduce crop yields.

It was this need to control ambient ozone formation, air toxics, acid raid, and stratospheric ozone depletion that led to the passage of the 1990 Clean Air Act. It should have far reaching effects. William K. Reilly, Administrator for the Environmental Protection Agency, claims that "This legislation will remove 56 billion pounds of pollution each year from the air we breath. That is 224 pounds of pollutants for every man, woman and child in this country. President Bush summed up the Act as being "...the most significant air pollution legislation in our nation's history."(3).

1990 Clean Air Act

The 1990 Clean Air Act sets emission standards, compliance dates, and enforcement provisions through ten sections (titles). Table I shows that Title II covers the control of pollutants that come from mobile sources, the subject matter for the first section of this book, while Titles I, III, IV, V, and VII cover the control of pollutants that could come from stationary sources, the subject matter for this chapter. The following synopses of these latter titles concentrate on the control requirements for the stationary sources.

Title I - Nonattainment. Title I addresses the pollution problems of ozone (smog), CO, and particulate matter (PM-10) by defining, for each of these pollutants, the geographical areas in the United States whose levels do not meet

TABLE I
CLEAN AIR ACT AMENDMENTS OF 1990

TITLE I: National Ambient Air Quality Standards
TITLE II: Mobil Sources
TITLE III: Air Toxics
TITLE IV: Acid Rain
TITLE V: Permits
TITLE VI: Stratospheric Ozone
TITLE VII: Enforcement
TITLE VIII: Miscellaneous
TITLE IX: Clean Air Research
TITLE X: Employment Transition

SOURCE: Adapted from reference 2.

their air quality standards, the required effectiveness of the control systems that must be used in these "nonattainment" areas, and the time tables for the non-attainment areas to reach their standards.

Table II shows that the ozone non-attainment areas are divided into the categories of Extreme (Los Angeles), Severe 1 (five areas), Severe 2 (four areas), Serious (fourteen areas), Moderate (thirty-one cities), and Marginal (forty-three areas). Also shown are the dates by which the cities in each of these categories are to attain their air quality standards. The areas with the worse problems have more time to do this. For example, cities in the marginal category have three years, while Los Angeles, in its extreme category, has twenty years.

These attainment periods for ozone are expected to be met through corresponding schedules for VOC reductions. Areas rated moderate and worse than moderate must reduce their VOC emissions by at least 15% within the first six years. Areas rated worse than moderate must then average a 3% VOC reduction per year for the remaining years in their attainment periods.

These VOC reductions are to be achieved through the implementation of a permit program, and the use of control technology. Existing stationary sources will require reasonably available control technology (RACT), and new stationary sources will require the technology that will achieve the lowest achievable emissions rate (LAER). RACT is to be defined by the EPA in their control technique guidelines (CTG). The CTGs, in turn, are to be adopted and implemented by the states as part of their state implementation plans (SIP).

The degree to which RACTs are to be applied depends on the category of the nonattainment areas. In all areas the existing RACT rules will be updated. Areas rated moderate and worse must apply the new RACT levels to existing and new CTGs, as well as to major non-CTG stationary sources.

The minimum sizes of these major non-CTG stationary sources are also defined by the categories of the areas in which they are operating. These sizes are 100, 50, 25 and 10 tons of VOCs/year for the moderate, serious, severe, and extreme areas, respectively. Major sources of NOx are mandated to meet the same control requirements as are the major sources of VOCs, unless the EPA concludes from its studies that such controls will have no benefit in controlling smog.

Title I handles CO in a manner similar to that for VOCs. The forty-two nonattainment areas are grouped into "moderate" and "serious" categories with five and ten year attainment dates, respectively. Control guidelines for stationary sources, which account for only 10% of the CO emissions, have not yet been issued.

Title III - Air Toxics. Title III has raised the number of air toxics to be controlled from the 7 that have been identified through the earlier identification process to 189 (Table III), and has set in place measures to insure their significant reduction within ten years. These measures, similar to the Title I VOC measures, define the sources and set their control standards.

These sources are classified as "Major Sources" and "Area Sources". Major sources are units that emit or have the potential to emit at least 10 tons per year of any one of the listed toxics, or 25 tons per year of any combination of them. Area sources are simply all of the remaining, smaller sources.

The degree to which a toxic leaving a major source needs to be controlled depends on the industry that the major source is serving. EPA has classified these industries into hundreds of categories or subcategories. Each of these categories will have an emission level based on its own maximum achievable control technology (MACT) for existing and new major sources.

The MACT for existing major sources is to be equivalent to that achieved by the best performing 12% of the existing major sources within a category with 30 or more sources. If a category has fewer than 30 major sources, MACT is to be based on its best performing five sources. Sources with recent LAER technology are not to be included in these determinations.

The MACT for new major sources is to be equivalent to that achieved by the best controlled similar source within each category. This standard is equivalent to a LAER standard.

EPA must promulgate all of these MACT standards within ten years. Their schedule calls for the 40 most serious major source categories and coke ovens to be regulated within two years, and 25%, 25%, and 50% of the remaining major source categories to be regulated within four, seven, and ten years, respectively.

Title III does not consider these steps sufficient by themselves; it goes on to provide for a second pass at the problem. Within eight years after the promulgation of the MACT standards for each major source category, EPA is to determine if the "residual risk" from the allowed emissions is sufficient to raise the cancer risk of the most exposed individual to one chance in a million. If it does, a new, adequately low standard will have to be met.

The EPA must also address the emissions from area sources. They need to develop a strategy to reduce the incidence of cancer from these emissions by 75%. Towards that end, they are to regulate area sources to the degree necessary to attain a 90% reduction of the thirty worse area source toxics. This

TABLE II

CLASSIFIED OZONE NONATTAINMENT AREAS
(Listed by Classification)

EXTREME
L.A.-S. Coast Air Basin CA

SEVERE 17

Chicago-Gary-Lake Cnty, IL-IN
Houston-Galveston-Brazoria, TX
Milwaukee-Racine, WI

New York-N New Jersey-
Long Island NY-NJ-CT
SE Desert Modified AQMA, CA

SEVERE 15

Baltimore,]D
Phil-Wilm-Trent, PA-NJ-DE-MD

San Diego, CA
Ventura Co, CA

SERIOUS

Altanta, GA
Baton Rouge, LA
Beaumont-Port Arthur, TX
Boston-Lawrence-Worcester
  (E.MA), MA-NH
El Paso, TX
Greater Connecticut
Muskegon, MI

Portsmouth-Dover-Rochester, NH
Providence (All RI), RI
Sacramento Metro, CA
San Joaquin Valley, CA
Sheboygan, WI
Springfield (W MA), MA
Washington, DC-MD-VA

MODERATE

Atlantic City, NJ
Charleston, WV
Charlotte-Gastonia, NC
Cincinnati-Hamilton, OH-KY
Cleveland-Akron-Lorain, OH
Dallas-Fort Worth, TX
Dayton-Springfield, OH
Detroit-Ann Arbor, MI
Grand Rapids, MI
Greensboro-Winston Salem-
  High Point, NC
Huntington-Ashland, WV-KY
Kewaunee Co, WI
Knox & Lincoln Cos, ME
Lewiston-Auburn, ME
Louisville, KY-IN

Manitowoc Co, WI
Miami-Ft.Ldale-W.Plm.Bch, FL
Monterey Bay, CA
Nashville, TN
Parkersburg, WV
Phoenix, AZ
Pittsburgh-Beaver Valley, PA
Portland, ME
Raleigh-Durham, NC
Reading, PA
Richmond-Petersburg, VA
Salt Lake City, UT
San Francisco-Bay Area, CA
Santa Barbara-Santa Maria-
  Lompoc, CA
St. Louis, MO-IL
Toledo, OH

Table II. Continued

MARGINAL

Albany-Schenectady-Troy, NY
Allentown-Bethlehem-Easton,
  PA-NJ
Altoona, PA
Birmingham, AL
Buffalo-Niagara Falls, NY
Canton, OH
Cherokee Co, SC
Columbus, OH
Door Co, WI
Edmonson Co, KY
Erie, PA
Essex Co (Whiteface Mtn), NY
Evansville, IN
Greenbrier Co, WV
Hancock & Waldo Cos, ME
Harrisburg-Lebanon-Carlisle, PA
Indianapolis, IN
Jeffercon Co, NY
Jersey Co, IL
Johnstown, PA
Kent and Queen Anne's Cos, MD

Knoxville, TN
Lake Charles, LA
Lancaster, PA
Lexington-Fayette, KY
Manchester, NH
Memphis, TN
Norfolk-Vir.Beach-Newport
  News, VA
Owensboro, KY
Paducah, KY
Portland-Vancouver AQMA, OR-
WA
Poughkeepsie, NY
Reno, NV
Scranton-Wilkes-Barre, PA
Seattle-Tacoma, WA
Smyth Co, VA (White Top Mtn)
South Bend-Elkhart, IN
Sussex Co, DE
Tampa-St. Petersburg-
  Clearwater, FL
Walworth Co, WI
York, PA
Youngstown-Warren-Sharon, OH-
PA

SUBMARGINAL
Kansas City, MO-KS

SOURCE:Adapted from reference 2

TABLE III

HAZARDOUS AIR POLLUTANTS

CAS #	Chemical Name	CAS #	Chemical Name	CAS #	Chemical Name
75070	Acetaldehyde	68122	Dimethyl formamide	82688	Pentachloronitrobenzene (Quintobenzene)
60355	Acetamide	57147	1,1-Dimethyl hydrazine	87865	Pentachlorophenol
75058	Acetonitrile	131113	Dimethyl phthalate	108952	Phenol
98862	Acetophenone	77781	Dimethyl sulfate	106503	p-Phenylenediamine
53963	2-Acetylaminofluorene	534521	4,6-Dinitro-o-cresol, and salts	75445	Phosgene
107028	Acrolein	51285	2,4-Dinitrophenol	7803512	Phosphine
79061	Acrylamide	121142	2,4-Dinitrotoluene	7723140	Phosphorus
79107	Acrylic acid	123911	1,4-Dioxane (1,4-Diethyleneoxide)	8549	Phthalic anhydride
107131	Acrylonitrile	122667	1,2-Diphenylhydrazine	1336363	Polychlorinated biphenyls (Aroclors)
107051	Allyl chloride	106898	Epichlorohydrin (1-Chloro-2,3-epoxypropane)	1120714	1,3-Propane sultone
92671	4-Aminobiphenyl	106887	1,2-Epoxybutane	57578	beta-Propiolactone
62533	Aniline	140885	Ethyl acrylate	123386	Propionaldehyde
90040	o-Anisidine	100414	Ethyl benzene	114261	Propoxur (Baygon)
1332214	Asbestos	51796	Ethyl carbamate (Urethane)	78875	Propylene dichloride (1,2-Dichloropropane)
71432	Benzene (including from gasoline)	75003	Ethyl chloride (Chloroethane)	75569	Propylene oxide
92875	Benzidine	106934	Ethylene dibromide (Dibromoethane)	75558	1,2-Propylenimine (2-Methyl aziridine)
98077	Benzotrichloride	107062	Ethylene dichloride (1,2-Dichloroethane)	91225	Quinoline
100447	Benzyl chloride	107211	Ethylene glycol	106514	Quinone
92524	Biphenyl	151564	Ethylene imine (Aziridine)	100425	Styrene
117817	Bis(2-ethylhexyl) phthalate (DEHP)	75218	Ethylene oxide	96093	Styrene oxide
542881	Bis(chloromethyl)ether	96457	Ethylene thiourea	1746016	2,3,7,8-Tetrachlorodibenzo-p-dioxin
75252	Bromoform	75343	Ethylidene dichloride (1,1-Dichloroethane)	79345	1,1,2,2-Tetrachloroethane
106990	1,3-Butadiene	50000	Formaldehyde	127184	Tetrachloroethylene (Perchloroethylene)
156627	Calcium cyanamide	76448	Heptachlor	7550450	Titanium tetrachloride
105602	Caprolactam	118741	Hexachlorobenzene	108883	Toluene
133062	Captan	87683	Hexachlorobutadiene	95807	2,4-Toluene diamine
63252	Carbaryl	77474	Hexachlorocyclopentadiene	584849	2,4-Toluene diisocyanate
75150	Carbon disulfide	67721	Hexachloroethane	95534	o-Toluidine
56235	Carbon tetrachloride	822060	Hexamethylene-1,6-diisocyanate	8001352	Toxaphene (chlorinated camphene)
463581	Carbonyl sulfide	680319	Hexamethylphosphoramide	120821	1,2,4-Trichlorobenzene
120809	Catechol	110543	Hexane	79005	1,1,2-Trichloroethane
133904	Chloramben	302012	Hydrazine	79016	Trichloroethylene
57749	Chlordane	7647010	Hydrochloric acid	95954	2,4,5-Trichlorophenol
7782505	Chlorine	7664393	Hydrogen fluoride (hydrofluoric acid)	88062	2,4,6-Trichlorophenol
79118	Chloroacetic acid	123319	Hydroquinone		
532274	2-Chloroacetophenone				
108907	Chlorobenzene				

510156	Chlorobenzilate	
67663	Chloroform	
107302	Chloromethyl methyl ether	
126998	Chloroprene	
1319773	Cresols/Cresylic acid (isomers and mixture)	
95487	o-Cresol	
108394	m-Cresol	
106445	p-Cresol	
98828	Cumene	
94757	2,4-D, salts and esters	
354704	DDE	
334883	Diazomethane	
132649	Dibenzofurans	
96128	1,2-Dibromo-3-chloropropane	
84742	Dibutylphthalate	
106467	1,4-Dichlorobenzene(p)	
91941	3,3-Dichlorobenzidine	
111444	Dichloroethyl ether (Bis(2-chloroethyl)ether)	
542756	1,3-Dichloropropene	
62737	Dichlorvos	
111422	Diethanolamine	
121697	N,N-Diethyl aniline (N,N-Dimethylaniline)	
64675	Diethyl sulfate	
119904	3,3-Dimethoxybenzidine	
60117	Dimethyl aminoazobenzene	
119937	3,3-Dimethyl benzidine	
79447	Dimethyl carbamoyl chloride	

78591	Isophorone
58899	Lindane (all isomers)
108316	Maleic anhydride
67561	Methanol
72435	Methoxychlor
74839	Methyl bromide (Bromomethane)
74873	Methyl chloride (Chloromethane)
71556	Methyl chloroform (1,1,1-Trichloroethane)
78933	Methyl ethyl ketone (2-Butanone)
60344	Methyl hydrazine
74884	Methyl iodide (Iodomethane)
108101	Methyl isobutyl ketone (Hexone)
624839	Methyl isocyanate
80626	Methyl methacrylate
1634044	Methyl tert butyl ether
101144	4,4-Methylene bis(2-chloroaniline)
75092	Methylene chloride (Dichloromethane)
101688	Methylene diphenyl diisocyanate (MDI)
101779	4,4'-Methylenedianiline
91203	Naphthalene
98953	Nitrobenzene
92933	4-Nitrobiphenyl
100027	4-Nitrophenol
79469	2-Nitropropane
684935	N-Nitroso-N-methylurea
62759	N-Nitrosodimethylamine
59892	N-Nitrosomorpholine
56382	Parathion

121448	Triethylamine
1582098	Trifluralin
540841	2,2,4-Trimethylpentane
108054	Vinyl acetate
593602	Vinyl bromide
75014	Vinyl chloride
75354	Vinylidene chloride (1,1-Dichloroethylene)
1330207	Xylenes (isomers and mixture)
95476	o-Xylenes
108383	m-Xylenes
106423	p-Xylenes
0	Antimony compounds
0	Arsenic compounds (inorganic including arsine)
0	Beryllium compounds
0	Cadmium compounds
0	Chromium compounds
0	Cobalt compounds
0	Coke oven emissions
0	Cyanide compounds [1]
0	Glycol ethers [2]
0	Lead compounds
0	Manganese compounds
0	Mercury compounds
0	Mineral fibers [3]
0	Nickel compounds
0	Polycylic organic matter [4]
0	Radionuclides (including radon) [5]
0	Selenium compounds

Note: For all listings above that contain the word "compounds" and for glycol ethers, the following applies: Unless otherwise specified, these listings are defined as including any unique chemical substance that contains the named chemical (i.e., antimony, arsenic, etc.) as part of that chemical's infrastructure.

(1) X'CN where X = H' or any other group where a formal dissociation may occur. For example, KCN or Ca (CN)2.

(2) Includes mono- and diethers of ethylene glycol, diethylene glycol, and triethylene glycol R-(OCH2CH2)n-OH or where n = 1,2 or 3: R = alkyl or aryl groups; R = R. H. or groups which, when removed, yield glycol ethers with the structure R-(OCH2CH2)n-OH. Polymers are excluded from the glycol category.

(3) Includes glass microfibers, glass wool fibers, rock wool fibers, and slag wool fibers, each characterized as "respirable" (fiber diameter less than 3.5 micrometers) and possessing an aspect ratio greater than or equal to 3, as emitted from production of fiber and fiber products.

(4) Includes organic compounds with more than one benzene ring, and which have a boiling point greater than or equal to 100 C.

(5) A type of atom which spontaneously undergoes radioactive decay.

SOURCE: Adapted from Reference 2.

will entail the use of standards that can be met from a Generally Available Control Technology (GACT).

Title IV - Acid Rain. The purpose of Title IV is to reduce, over this decade, SO_2 emissions by approximately ten million tons/year and NOx emissions by approximately two million tons/year. Most of these reductions will come from the utilities and will result in a 8.9 million tons/year cap for SO_2 by the year 2000.

SO_2 reductions from utilities are to be attained through an innovative market-based system that allows utilities to buy and sell allowances, which are rights to emit one ton of SO_2/year. The utilities would be allowed to use any control techniques they desired to keep their actual emissions within their held allowances. Emissions from new powerplants would have to be offset by corresponding reductions in their older plants, or be matched by newly purchased allowances.

Although EPA is to study the plausibility of trading SO_2 allowances for NOx allowances, present NOx control measures are to come from technology requirements. EPA is to set NOx standards for some existing utility boilers within the next year, and for all existing utility boilers by 1997. Furthermore, they are to establish new source performance standards (NSPS) by 1993.

Title V - Permits. The comprehensive permit system established under Title V is considered to be the most important procedural reform in the new Clean Air Act. It, along with the penalty provisions under Title VII, will enhance EPA's ability to enforce the technical control requirements defined for the various stationary sources described under Titles I, III, and IV.

The EPA will set the guidelines for the state permit programs, review these state programs, and oversee their implementation. The permits themselves will incorporate all of the obligations required of the stationary source, as well as include a comprehensive compliance schedule with applicable monitoring and reporting requirements.

Title VI - Enforcement. This title has a broad range of powers to make the Clean Air Act more enforceable. EPA can now issue administrative penalties up to $200,000 and field citations up to $5,000. Owners must certify their compliances, with EPA having a right to subpoena supporting data. Criminal penalties are upgraded from misdemeanors to felonies, and individual responsibility is assigned to senior management and corporate official levels. Citizens may now seek penalties against violators, with the penalties going towards EPA's enforcement activities.

With these national standards, technological controls, permitting system and enforcement provisions, the 1990 Clean Air Act is a comprehensive act. However, no summary of pollution legislation is complete without mentioning state legislation, particularly that of California, which is expected to be the pacesetter for many of the forthcoming state programs required by the 1990 Clean Air Act.

California Clean Air Act

California's 1988 Clean Air Act was written to correct their unusually severe pollution problems, which were not being addressed by the 1977 federal act. It too defined air quality standards, designated attainment and non-attainment areas, divided areas of nonattainment into categories of severity, stipulated reasonably available control technology (RACT), best available control technology (BACT), and best available retrofit control technology (BARCT) as corrective measures, established a permit system financed by fees, and established civil penalties.

It is California's control technology standards that will undoubtedly affect the 1990 federal act the most. The requirements for motor vehicles are already being embraced by 11 northeastern states. It is expected that California's RACT and BACT requirements will influence the RACT, MACT, and LAER requirements of the federal act.

These different control requirements will emanate primarily from the control technologies that are widely practiced in industry today. The preferred choice of one technology over the others will often depend on the market value and the physical properties of the pollutant.

VOC and Toxic Control Technologies

General. VOC's and toxics are usually controlled by carbon adsorption, condensing systems, wet scrubbing, and incineration. Incineration can be either thermal or catalytic.

Carbon adsorption systems remove emissions with a packed bed of carbon, which generally has a high adsorption preference for hydrocarbons. These adsorbed compounds are subsequently stripped from the bed with hot air or steam. The higher temperature of this stripping gas allows its quantity to be less than the quantity of the contaminated stream from which the hydrocarbons were initially removed. The resulting higher hydrocarbon concentration in the stripping gas sometimes allows the hydrocarbon to be recovered with a simple additional process step. Such recovery can make carbon adsorption an attractive control technology for expensive compounds whose worth can be credited against the cost of the adsorption process.

Condensing systems remove emissions with a chiller and a knock-out pot. The temperature of the coolant used in the chiller must be at a temperature that is above the melting point of the targeted compound, but still low enough to insure that the vapor pressure of the compound's condensed phase is low enough for the stream to meet its allowable emission level. Since the refrigeration of the coolant can be expensive, condensing systems are most attractive when the recovered compound is valuable.

Wet scrubbing systems remove water soluble emissions from an up-flowing stream in a bed that is packed with inert packing and is wetted with a down-flowing aqueous stream. The aqueous stream may contain chemicals that react with the absorbed compound. While such scrubbing systems are inexpensive and can achieve very high emission reductions, they sometimes only change an air emissions problem into a waste water problem.

Incineration systems remove emissions by oxidizing them to their basic oxidation products, usually CO_2 and H_2O in either a large high temperature chamber (thermal incineration) or in a smaller packed bed of catalyst (catalytic incineration). Both systems can destroy emissions cleanly with 99+% conversions.

It is this clean, complete destruction that is expected to make incineration systems the ideal choices for the LAER, RACT, and MACT technologies required by the Clean Air Act. In fact, these destruction characteristics will probably make incineration systems ideal choices for the final cleaning up of the streams with unacceptably high contaminant levels leaving carbon adsorption and condensing systems.

VOC Control. The manufacture of many consumer products involve the use of organic compounds. During processing, small quantities of these VOC's may be released into the atmosphere. Table IV shows that there is a large diversity of industries that are doing this. If they are located in an nonattainment area, they will have to control these emissions to the degree specified under Title I of the 1990 Clean Air Act.

TABLE IV
SOME OF THE APPLICATIONS FOR THE CONTROL OF VOLATILE
ORGANIC COMPOUNDS

• Diesel trucks	• Ovens
• Bakeries	• Painting
• Breweries	• Dry Cleaning
• Chemical plants	• Paper
• Coating processes	• Petrochemicals
• Electronics industry	• Petroleum Storage
• Food Processing	• Pharmaceutical
• Furniture manufacture	• Power Generation
• Groundwater Cleanup	• Printing
• Hazardous Waste	• Wood Coating
• Landfills	• Stationary Engines

SOURCE: Adapted from reference 14.

The preferred technology for VOC control is incineration. The economic choice, particularly for moderately sized streams with moderate amounts of VOCs, is catalytic incineration. The primary reason for this is the energy cost.

Catalytic incineration typically starts in the 300-500°F range, whereas the onset of thermal incineration is in the 800-1100°F range. These temperatures are typically increased to obtain maximum performance, with the catalytic units requiring inlet temperatures ranging from 600-80 °F and the thermal units requiring inlet temperatures ranging from 1200-1800°F. The halving of fuel requirements for the catalytic incineration relative to thermal incineration units are further enhanced with an inlet/outlet heat exchanger, which can lower the fuel requirements to just the minimum amount necessary for stable process control.

Since NO_x formation is directly proportional to the quantity of fuel (with its attendant high temperature burning zone) consumed, catalytic units also produce far less NO_x than do thermal units.   The same is true for CO_2 production, a major contributor to the greenhouse effect. Finally, catalytic units have no problem oxidizing CO at their usual operating temperatures, whereas thermal units must operate above 1600°F to destroy CO.

The effective design of a catalyst system for a particular VOC control application starts with the definition of the required conversion level of the system over its lifetime. This level will be defined by the states' RACT or LAER requirements. It may be obtained through the adjustment of the inlet temperature, the volume of the catalyst bed, the amount of geometric surface area available in a given volume within the catalyst bed, and the quantity and type of precious metal on the catalyst. Optimization of these parameters routinely achieves 90-99% destruction efficiencies, with some reaching 99.9 + %.

In many other situations, the primary mode of deactivation of VOC control catalysts is poisoning. The phenomena has been well documented(5). For these applications, catalyst design criteria are used to minimize poisoning-(6). These include the use of guard beds, less poison susceptible noble metals (e.g., platinum instead of palladium), and appropriate substrate and catalyst bed configurations.

The general construction and design of catalysts have been described by DeLuca and Campbell(7), while the oxidation of a wide range of VOCs, both as pure components and as mixtures, has been described by Tichenor and Palazzo(8).

The work in this book includes an investigation of the equilibrium between metal ion complexes and oxide surfaces, an important factor in the manufacture of commercial VOC catalysts, and the catalytic photodegradation of furans.

Toxic Control. The preferred control technologies for toxics are not yet easy to discern, since the problem is just beginning to be addressed. However, it can be expected that catalytic and thermal incineration will play a major role in controlling toxics.

An important subclass of toxics are those that contain halogen atoms. They are generated from a variety of applications, including the manufacture of organic chemicals (e.g., vinyl chloride monomer), the production of polymers, and the degreasing operations used in metal processing. In addition, the vent from some air stripping units to clean groundwater or contaminated soil contain halogenated hydrocarbons.

As a class, halogenated hydrocarbons are among the more difficult compounds to destroy by thermal incineration. Similarly, the noble metals usually used in VOC incineration catalysts can be severely inhibited by the halogen atoms from the halogenated hydrocarbon.

Past attempts to develop catalysts that are not affected by halogens have been partially effective. At least two processes have been handling halogenated organics with fluidized catalyst beds, one of which is believed to have destroyed over twenty million pounds per year of chlorinated by-products from a vinyl chloride monomer plant(9,10).

The need for continual catalyst replacement, fines entrapment, and larger beds for comparable conversions makes fluidized beds a poor alternative to fixed packed beds. Consequently, considerable effort has gone into the development of a fixed bed catalyst that can handle halogenated organics.

Some of these efforts have already led to a catalyst(11) that has shown a capability to destroy a number of diverse chlorinated and fluorinated hydrocarbons with more activity and stability than the supported noble metals or chromia-alumina catalysts used previously.

The work in this book describes the activities of commercially common Pt/Al_2O_3 and Pd/Al_2O_3 monolithic catalysts for the oxidative destruction of the well-known contaminant in groundwater, trichloroethlyene. It also presents a more fundamental surface study on the thermal decomposition of alkyl halides on copper surfaces.

NOx Control Technologies

General. Nitrogen and oxygen exist as a variety of stable, but inter-convertible, oxides. They are known collectively as NOx. Nitrogen dioxide (NO_2) has already been described as an ozone precursor. It, along with nitric oxide (NO), also causes respiratory problems over long periods of exposure. While the more abundant nitrous oxide (N_2O) is not directly hazardous to human health, it is a "greenhouse gas" that may contribute to global warming. It also is one of the gases that destroys stratospheric ozone. While most of the NO_2 and NO present in the atmosphere is manmade, most of the N_2O comes from biochemical reactions in the soil.

Two general types of catalytic processes can be used to control NOx emissions. The process used depends on the gas stoichiometry entering the catalyst bed. If the gas has an excess or equivalent amount of reductants compared to the amount of oxidants present, NOx can be controlled by the use of conventional automotive-type precious metal catalysts. This type of NOx control process is referred to as non-selective catalytic reduction (NSCR).

If the catalyst is exposed to an exhaust steam that is net oxidizing, a second type of process is used. This process is referred to as selective catalytic reduction (SCR). During selective catalytic reduction, a reducing agent, most commonly ammonia, is injected into the NOx containing exhaust. The ammonia and NOx react over the SCR catalyst to form nitrogen and water.

SCR catalysts typically are composed of base metal compounds. Precious metal catalysts usually have much smaller temperature and inlet composition windows than do commercial base metal catalysts. Consequently, precious metal catalysts are generally not used for the SCR application.

NSCR Process. When the exhaust stream composition is close to stoichiometric, rhodium-containing precious metal catalysts can be used to control NOx emissions. The automobile industry has used such a "three-way" catalyst for years to simultaneously convert the NOx, CO and hydrocarbons in automotive exhausts into water, CO_2, and N_2. This feat requires that the air-fuel ratio in the gas entering the engine to be held very near the stoichiometric point. Keeping this ratio steady under widely different driving conditions has necessitated the development of a feed-back control system that uses the oxygen content of the engine exhaust as the control signal to adjust the incoming air or fuel streams.

The NOx, CO and hydrocarbon emissions from stationary internal combustion engines (powering electric generators, pumps, and compressors) can be controlled in the same manner (Figure 1). The relatively constant loads of these engines may appear to make the control of their air/fuel ratio easier than is the case for conventional automotive engines. However, most of the stationary internal combustion engines are run on natural gas(12), and the air/fuel ratio requirements for controlling these emissions are more stringent (13).

Automotive three-way catalysts have been designed to withstand several thousand hours of use in an environment where they are continuously exposed to poisoning from lubricating oils and high temperatures that can cause irreversible thermal damage to the noble metals through sintering and alloy formation. This built-in protection has proven to be more than adequate for the three-way catalysts serving natural gas fueled engines, with their lower contaminant levels and their relatively stable exhaust conditions.

While the work reported in this book on NSCR catalysts is in the section on automotive catalysts, most of it is germane to stationary internal combustion engines. This is particularly true for the work done on methane oxidation, on the CO oxidation reactions, on the effect of the catalyst support, on the effect of ceria, and on the relative ease of destruction for various hydrocarbon compounds on a three-way catalyst.

SCR Process. Selective Catalytic Reduction technology is widely used in both Japan and Europe to reduce NOx emissions formed during power generation. The most commonly used catalysts contain titania and vanadia. The vent stream from coal or oil fired boilers is strongly oxidizing -- an operational condition in which rhodium-containing three-way catalysts give low conversions of NOx.

The central catalytic event behind the successful operation of selective catalytic reduction systems is that under lean operation NOx reacts with ammonia, or a nitrogen containing base such as urea, rather than with the excess oxygen in the flue stream, according to the following reaction:

$$3\,NO + 2\,NH_3 \rightarrow 5/2\,N_2 + 3\,H_2O$$

The process requires good control of the ammonia injection rate. An inadequate injection rate results in unacceptably low NOx conversions. A too high injection rate results in the venting of undesirable ammonia to the atmosphere.

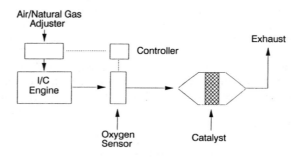

Figure 1. Schematic of a three-way catalytic control system with natural gas fuel. (Adapted from reference 14.)

The work done on SCR catalysts in this book covers the effects of the structural characteristics of both bulk and supported vanadia on their activities and selectivities.

Conclusions

This chapter has described the main features of the 1990 Clean Air Act that pertain to the controlling of emissions from stationary sources, has briefly reviewed the primary types of control technologies that are generally used on stationary sources, has emphasized where catalytic incineration has gained prominence, and has lightly touched on the recent status of the catalyst technologies for these applications.

It is apparent from this review that catalytic incineration, with its inherently low fuel costs, will have a large role to play in the meeting of the requirements of the 1990 Clean Air Act, particularly in controlling VOCs and air toxics. This blossoming demand can be expected to lead to many exciting developments in catalyst surface chemistry over the next decade.

Acknowledgement

The authors thank R. C. Conner, Technical Director of the Manufacturers of Emission Controls Association, for his help in summarizing the provisions of the Federal and California clean air acts.

Literature Cited

1) S. L. S. Dombrowski, "The Revised Clean Air Act: What Does It Mean and What Will it Cost?", *Environmental Protection*, April-May (1991).
2) "The Clean Air Act Amendments of 1990 Summary Materials", U.S. EPA (November 15, 1990).
3) W. G. Rosenberg, *Automotive News - World Congress Conference*, Detroit, MI, January 14, 1991.
4) "National Air Quality and Emissions Trends Report, 1989", EPA-450/4-91-003 (February 1991).
5) G.R. Lester and J.C. Summers, "Poison-Resistant Catalysts for Purification of Web Offset Press Exhaust", Paper No. 88-83.7.
6) L. L. Hegedus and R.W. McCabe, "Catalyst Poisoning", *Chemical Reviews Science Engineering*, 23(3), 377, 1981.
7) J. P. DeLuca and L.E. Campbell, in *"Monolithic Catalyst Supports"*, J. J. Burton and R. L. Garten; Advanced Materials in Catalysis, Academic Press, Chapter 10, p293 - 324.
8) B.A. Tichenor and M.A. Pallazolo. "Destruction of Volatile Organic Compounds Via Catalytic Incinerator", *Paper No. 49B, AIChE 1985 Annual Meeting*, November 10-15, Chicago (1985).
9) L. C. Hardison and E. J. Dowd, "Emission Control Via Fluidized Bed Oxidation", *Chem. Eng. Prog.* 73 (7): 31 (1977).

10)   J. S. Benson, "Catoxid for Chlorinated By-Products", *Hydrocarbon Proc.* 58 (10): 107 (1979).

11)   G. R. Lester,"Catalytic Destruction of Hazardous Halogenated Organic Compounds", Paper No. 89.96A.3.

12)   "Proposed Strategy for the Control of Oxides of Nitrogen from Stationary Internal Combustion Engines", State of California Air Resources Board (October 1979).

13)   J. Klimstra, "Cataytic Converters for Natural Gas Fueled Engines - A Measurement and Control Problem", *SAE Paper 872165* (1987).

14)   J. C. Summers and A. C. Frost in *Proceedings of the 1991 Platinum Group Metals Seminar*; N. Carson, M.I.E. Guindy, and J. Luben, eds.; International Precious Metals Institute: Austin, Texas, 1991.

RECEIVED March 20, 1992

Chapter 9

Selective Catalytic Reduction of Nitric Oxide with Ammonia over Supported and Unsupported Vanadia Catalysts

U. S. Ozkan, Y. Cai, and M. W. Kumthekar

Department of Chemical Engineering, The Ohio State University, Columbus, OH 43210

Vanadia catalysts supported over titania and unsupported V_2O_5 catalysts were investigated for selective catalytic reduction of NO with NH_3. All catalysts were characterized using BET surface area measurement, X-ray diffraction, in-situ and ambient laser Raman spectroscopy, X-ray photoelectron spectroscopy, and scanning electron microscopy techniques. A steady-state plug-flow reactor was used to investigate the catalyst-support interactions and the structural specificity of V_2O_5 in selective catalytic reduction of NO through activity and selectivity measurements. The activity of the catalysts supported over both anatase and rutile phases went through a maximum with vanadia loading. Laser Raman studies for supported catalysts suggested the presence of polymeric vanadate species at low loading levels. For bulk V_2O_5 catalysts, preferential exposure of different crystal planes was found to change the activity and product distribution significantly.

Pollution of the environment by nitrogen oxide emissions is a rapidly growing concern. These emissions mainly result from burning of fossil fuels and bio-mass. The hazards associated with nitrogen oxides include respiratory health problems in human beings, a direct phytotoxic effect on the plant communities, increase in "green-house" gases, and changes in ozone layer. Selective Catalytic Reduction (SCR) of the oxides of nitrogen has proven to be an effective method to reduce the NOx emissions from stationary combustion sources. In the recent past, several studies have been devoted to investigate the mechanism of the reaction, the interaction of the catalyst and the support and the effect of the reaction parameters on catalytic activity. Vanadia catalysts supported over titania have proven to be much superior to other catalysts because of their high activity and high resistance to poisoning by SO_2. A recent review article by Bosch and Janssen (1) has summarized most of the work done in this area. In spite of the extensive studies conducted over supported vanadia catalysts(2-4), the controversy as to the role of the support is still unresolved. There are conflicting reports about the relative activity of the two phases of the titania support, anatase and rutile.

The structural specificity of bulk V_2O_5 catalysts and the role played by different planes of the crystal in the mechanism of the reaction are other important questions that

0097–6156/92/0495–0115$06.00/0

need to be addressed to gain a better insight to the catalytic phenomena involved in SCR reactions over both supported and unsupported catalysts. Morphological aspects of V_2O_5 catalyst were studied by Gasior and Machej (5) in the oxidation of o-xylene. Adsorption studies over the V_2O_5 catalysts by Andersson (6) attempted to locate the V=O species on the crystal of V_2O_5. An attempt in a similar direction was made by Miyamoto et al. (7,8) to quantify the number of V=O sites on the surface of the catalyst.

In earlier articles, we have reported the results of our studies dealing with the effect of support material and reaction parameters on catalytic behavior (9, 10). In this article, we are reporting reactivity experiments for selective catalytic reduction of nitric oxide with ammonia performed over both supported and unsupported catalysts. Catalysts were prepared by supporting vanadia over both phases of titania and by growing V_2O_5 crystals with preferred orientation . All catalysts were characterized using BET surface area measurement, X-ray diffraction, scanning electron microscopy, 3-D imaging technique, X-ray photoelectron microscopy, and in-situ and ambient laser Raman spectroscopy. This study has aimed at understanding the effect of the support crystal structure on the activity of the vanadia surface species and the morphological aspects in reduction of NO over V_2O_5 catalysts.

Methods

Catalyst Preparation.

The methods outlined previously in the literature (11, 12) were used to prepare supported vanadium catalysts over titania. The anatase phase of titania was obtained from Aldrich, while the rutile phase was prepared by heating a mixture of anatase and rutile phases (Degussa P25) to 875°C for 12 hours under the flow of oxygen. The surface areas of TiO_2 (Anatase) and TiO_2 (Rutile) were 10.5 and 6.2 m^2/g, respectively. Supports were wet-impregnated with a solution of ammonium metavanadate (Aldrich) in oxalic acid (Fisher) at 90°C for 5 hours, followed by drying at 65°C under vacuum and heating at 110°C in air for 10 hours. The catalysts were then calcined at 500°C for 4 hours under the flow of oxygen. The unsupported V_2O_5 catalysts were prepared in two different ways. Ammonium metavanadate (Aldrich) was calcined at 520°C for 50 hours in the flow of oxygen to give sample D. Pure V_2O_5 (99.6%) from Aldrich was melted at 695°C for 2 hours and then subjected to temperature programmed cooling for recrystallization to give sample M. The surface areas of samples D and M were 4.2 and 0.25 m^2/g, respectively. Both the supported and the unsupported catalysts were loaded to the reactor in powder form.

Catalyst Characterization.

The specific surface areas of supports, and of supported and unsupported catalysts were measured using the BET technique with a Micromeritics 2100E Accusorb instrument. Nitrogen and krypton were used as the adsorbents. X-ray powder diffraction patterns of the samples were obtained using a Scintag PAD V diffractometer. Cu K_α radiation (λ=1.5432 Å) was used as the incident X-ray source. The interaction of the vanadium oxide with the support materials was examined using a laser Raman spectrometer (SPEX 1403 Ramalog 9-I Spectrometer). A 5-W Argon ion laser (Spectra Physics, model 2016) was used as the excitation source. The in-situ Raman experiments were performed using a specially designed high-temperature, controlled-atmosphere cell. The surface morphology of the catalysts was examined through a Hitachi S-510 scanning electron microscope using a voltage of 25 KV and magnifications ranging from 100X to 10,000X. X-ray photoelectron spectra of the samples were obtained using a Physical Electronics/Perkin Elmer (Model 550) ESCA/Auger spectrometer, operated at 15 kV, 20 mA. The X-ray source was Mg K_α radiation (1253.6 eV). The binding energy for C 1s (284.6 ev) was used as a reference in these measurements.

Reaction Studies.

The reactor system used for selective catalytic reduction was described previously [10]. All tubing was 1/4" 304 stainless steel unless otherwise indicated. A steady-state fixed-bed reactor was used with the dimensions of 6.4 mm O.D., 4.6 mm I.D. The catalyst bed was about 2 cm long. An iron-constantan type J thermocouple (Omega) was placed at the center of the catalyst bed. The temperature was controlled and displayed with an Omega temperature controller (Model 4001.JC). The feed gases (Linde) consisted of 0.477% NO in He, 0.53% NH_3 in He, 10% O_2 in He or 1% O_2 in He, and pure He. The gas flow rates were adjusted using four mass flow controllers (Tylan model FC-280) and a four-channel readout box (Tylan model RO-28). All the exit lines from the reactor were heated to prevent condensation of water and formation of ammonium salt on the tubing walls.

Analyses of feed and product streams were performed by combining gas chromatography, chemiluminescence and titration techniques. Inlet and outlet concentrations of NO and NO_2 (if present) were obtained with an on-line chemiluminescence NO-NO_x analyzer (Thermo Environmental Instruments, Model 10) which includes a high temperature converter for NO_2 and NH_3 and a low temperature catalytic converter (Model 300) only for NO_2. Although the chemiluminescence analyzer is equipped with the capability for converting ammonia, our studies showed that the presence of ammonia interfered with NO and NO_2 measurements. Using a procedure outlined in the literature [11], ammonia was removed by absorbing it in concentrated (85%) phosphoric acid prior to the chemiluminescence analysis. The NH_3 concentrations in the feed and product streams were determined by a conventional titration method with the aid of hydrochloric acid solution of known concentration after NH_3 was absorbed in a diluted aqueous boric acid solution. An on-line gas chromatograph (Hewlett Packard 5890 A) with a 10-ft Porapak Q column and a 8-ft molecular sieve 5A column was used to quantify O_2, N_2 and N_2O concentrations. The two columns were connected in series by a four-port Valco column isolation valve.

Activity measurements were performed using a feed mixture that consisted of 1465 ppm NO, 1418 ppm NH_3, and 0.88 % oxygen with helium as the balance gas. The reaction temperatures were varied from 150 to 400°C. The total volumetric flow rate of 100 cm^3(STP)/min was maintained constant. The total surface area in the reactor was kept constant for all catalyst loadings for a given support material. The results obtained from kinetic experiments were reproducible within 5%.

Results

Catalyst Characterization.

When X-ray diffraction patterns of V_2O_5 catalysts supported over both rutile and anatase phases of titania were obtained, the presence of crystalline V_2O_5 was detected only at a loading level of 6.6% (wt.%) and higher. At lower loading levels, the only reflections observed were those of the support material. The X-ray diffraction results for the unsupported V_2O_5 gave a clear indication of the preferential exposure of different crystal planes. The ratio of relative intensities for various (hkl) reflections are shown in Table I. The band assignments are based on the orthorhombic system with unit cell dimensions of $a_0=11.51$ Å, $b_0=4.371$ Å, and $c_0=3.559$ Å. The X-ray diffraction pattern obtained for sample D corresponded to the data reported in JCPDS Files (#9-387).

Table I. Comparison of X-Ray Diffraction Data

Ratio of intensities	V_2O_5-M	V_2O_5-D
$\Sigma I_{(h00)}/I_{(010)}$	0.04	1.07
$\Sigma I_{(00l)}/I_{(010)}$	<0.01	0.26
$I_{(101)}/I_{(010)}$	<0.01	0.73

The scanning electron microscopy studies over the supported catalysts have shown the presence of V_2O_5 at high loading levels. A uniformity was observed in the dispersion of vanadia phase over both phases of TiO_2. The SEM studies for unsupported catalysts showed major difference between the two samples. Sample D consisted of thick and chunky crystals, while sample M consisted of thin, long, and sheet-like crystals.

The laser Raman spectra of vanadia catalysts supported on the anatase phase of titania have been reported previously (10). The major Raman bands of the anatase phase that appear at 138, 192, 309, 392, 511, 634 and 790 cm^{-1} are in agreement with those reported earlier in the literature (13-16). The crystalline V_2O_5 exhibits Raman bands at 100, 141, 193, 280, 300, 401, 483, 528, 702, 995 cm^{-1} which agree with the previously reported in the literature (13,14,17-20). The Raman bands corresponding to crystalline V_2O_5 are clearly visible at the higher loading levels. The indication of the presence of crystalline V_2O_5 is the Raman band at 995 cm^{-1} (20). The band corresponding to crystalline V_2O_5 is clearly visible at loading levels of 4.4% and higher. At loading levels of 2.3% and lower, this band does not exist. A close inspection of the band at 995 cm^{-1} reveals that, at loading levels of 2.3% and below, the sharp band of 995 cm^{-1} is replaced by a much broader band ranging from 940 to 1025 cm^{-1}. This band is believed to be associated with polyvanadate surface species (15, 21, 22). In-situ Raman experiments performed over the dehydrated samples did not give any sharp bands at 1030 cm^{-1}, which would be indicative of monomeric surface species. The fact that the surface area of the support used in this study is considerably lower than those reported in the literature can explain the presence of only the polymeric vanadate species on our samples. Also the fact that the broad band seen in 940-1025 cm^{-1} region is an indication of longer chain length for the polymeric surface species.

The Raman spectra of vanadia catalysts supported over the rutile phase have also been reported in a previous publication (*10*). The Raman bands for the rutile phase of titania appear at 138, 231, 443, 606, and 822 cm^{-1} which agree with those reported earlier by other researchers (*15,16, 23*). The 995 cm^{-1} band remains visible at loading levels as low as 1.1 %. A close inspection of the spectrum revealed that the sharp band at 995 cm^{-1} coexisted with a broad band ranging from 940 to 1025 cm^{-1} at loading levels of 2.3% and below. At loading levels of 0.7% the sharp band at 995 cm^{-1} disappeared completely.

The Raman spectra of unsupported catalysts (sample D and sample M) showed some differences in the relative intensities. The intensity ratios of some of the major bands are reported in Table II.

Table II. Comparison of Laser Raman Spectroscopy Data

Ratio of Intensities	V_2O_5-D	V_2O_5-M
$I_{(995 \, cm^{-1})}/I_{(702 cm^{-1})}$	2.11	2.30
$I_{(995 \, cm^{-1})}/I_{(528 cm^{-1})}$	1.97	2.31
$I_{(995 \, cm^{-1})}/I_{(483 cm^{-1})}$	3.13	3.52

The XPS binding energy values obtained for bare supports, for supported catalysts, and for crystalline V_2O_5 are shown in Table III. for supported catalysts, while the binding energies for titanium did not show any shifts or band broadenings when the support was impregnated with the vanadia. However, there were considerable differences between the vanadium binding energies of the bulk V_2O_5 and supported vanadia accompanied by substantial band broadening, indicating a change in the coordination environment of the vanadia species when coordinated to the surface. There were no major differences between the binding energy values or band widths for the two unsupported V_2O_5 samples (D and M).

Table III. Binding energies for vanadium and titanium obtained by XPS

Sample	V (2p$_{3/2}$) (eV)	Half width (eV)	Ti (2p$_{3/2}$) (eV)	Half width (eV)
TiO_2 (A)	-	-	458.2	1.80
TiO_2(R)	-	-	458.3	1.81
2.3%V_2O_5/TiO_2(A)	516.7	2.46	458.5	1.85
2.3%V_2O_5/TiO_2(R)	516.8	2.31	458.5	1.86
V_2O_5-D	517.3	1.95	-	-
V_2O_5-M	517.6	1.95	-	-

Reaction Studies.

Blank Reactor and Bare Support Runs. The stainless steel reactor and the quartz wool were tested for catalytic activity by performing blank reactor runs at temperatures varying from 200 to 400°C. The conversion of NO remained below 3% up to a temperature of 350 °C. At 400 °C, which is the upper limit for SCR applications, the conversion was only 8%. Ammonia oxidation, which results in formation of N_2O, remained at zero up to a temperature of 350°C. At temperatures around 400°C, approximately 10% of ammonia was seen to oxidize to N_2O. Blank runs were also performed over bare supports, both anatase and rutile, at 250°C. The maximum conversion rate obtained over the bare support was less than 0.007 micromol/g.sec. This level of activity is less than 2 % of the activity observed over the supported catalysts.

Effect of Temperature. The effect of temperature on the activity and selectivity of vanadia catalyst supported on anatase phase of titania was investigated by varying the temperature range between 100 to 400°C. The activity of the catalyst was found to be negligible at temperatures below 100°C whereas a substantial increase in the activity was found when the temperature was raised from 150 to 250°C. At 300°C the conversion of NO was maximum, while the conversion of NH_3 continued to increase, which can be attributed to the direct oxidation of NH_3. This explains the formation of N_2O at temperatures higher than 300°C and also the decrease in the conversion of NO because of the lack of NH_3.

Effect of Oxygen Concentration. The activity of the catalyst was found to be strongly dependent on the concentration of oxygen in the gas phase for reduction of NO with ammonia. This effect was investigated over 2.2% vanadia catalyst supported over the anatase phase of titania at a constant temperature of 250 °C. At oxygen concentration levels below 1%, the NO conversion rate increased sharply with increasing oxygen concentration. At O_2 concentration levels above 1%, this effect was negligible, activity not responding to further increases in the oxygen concentration.

Effect of Catalyst Loading and the Support Material. To investigate the effect of catalyst loading, the activity of the supported catalyst was measured as the function of catalyst loading. The temperature was kept constant at 250°C for all the values of catalyst loadings. The concentration of the inlet gases was kept the same as those used in the previous section. The activity measurements are summarized in Table IV. The activity was found to be maximum at a loading of 2.3% for the anatase phase of titania. A similar but broader maximum is obtained for the vanadia catalyst supported over rutile phase of titania. The activities of the catalysts over the two supports seem to be comparable although the catalysts supported over the rutile phase appear to be more active on a unit surface area basis.

Structural Specificity of V_2O_5 in SCR Reaction. To investigate the structural specificity of V_2O_5, samples which preferentially exposed different crystal planes (samples D and M) were subjected to reaction conditions over a temperature range of 150 to 400°C. The activity was measured in terms of the rate of consumption of NO based on unit surface area. At temperatures below 200°C both samples showed very low activity. Both samples showed maximum activity at 350°C, although sample M was much more active than sample D. The increase in the activity with temperature for sample M was much more pronounced, whereas it was comparatively slower for sample D. At lower temperatures (<250°C) sample D was more active than sample M, whereas at temperatures higher than 250°C sample M was found to be more active than sample D. The two samples also showed major differences in the ratio of NH_3 conversion rate to NO conversion rate and in product distribution. While Sample M, which preferentially exposed the basal (010) plane converted ammonia at a faster rate than it converted NO, it also gave larger ratios of N_2O yield to N_2 yield. Table V summarizes these comparisons.

Table IV.   The Rate of NO Conversion over Supported Vanadia Catalysts

Wt % Vanadia Content	Conversion Rate (μmol/m^2.sec) $*10^3$
1.1%V_2O_5/TiO_2(A)	17
2.3%V_2O_5/TiO_2(A)	29
4.4%V_2O_5/TiO_2(A)	22
6.6%V_2O_5/TiO_2(A)	19
11%V_2O_5/TiO_2(A)	18
0.7%V_2O_5/TiO_2(R)	25
1.1%V_2O_5/TiO_2(R)	40
2.3%V_2O_5/TiO_2(R)	42
4.4%V_2O_5/TiO_2(R)	38
6.6%V_2O_5/TiO_2(R)	34

Table V.   Comparison of Conversion Rates and Product Distributions over V_2O_5 Crystals with Preferentially Exposed Crystal Planes

Temperature	N_2O yield/N_2 yield		NH_3 con. rate/NO con. rate	
	V_2O_5-D	V_2O_5-M	V_2O_5-D	V_2O_5-M
300 °C	0.05	1.46	1.07	1.17
350 °C	0.14	1.15	1.13	2.89
400 °C	0.65	0.81	1.31	12.60

Discussion

The experimental studies performed over supported and unsupported vanadia catalyst reemphasize the fact that vanadia catalysts provide high activity levels with high selectivities towards N_2 at lower temperatures. Laser Raman spectroscopy studies showed the existence of coordinated polymeric surface species at and below the loading level of 2.3% for the anatase phase of the support. The activity studies as a function of the loading level complemented with the Raman spectroscopy studies indicate that the coordinated surface species are more active than crystalline V_2O_5. The decrease in the activity of the catalyst below the critical loading level can be explained by the decrease in the number of active sites and the exposure of the bare support which is inactive. For the rutile phase, crystalline V_2O_5 and coordinated polymeric species were shown to coexist by the laser Raman spectroscopy studies at loading levels of 1.1 and 2.3%. The kinetic experiments showed maximum activity at a loading level of 2.3% suggesting that the presence of crystalline V_2O_5 did not affect the activity of the catalyst significantly.

The scanning electron microscopy studies combined with 3-D imaging technique, which we described in a previous publication (24), allowed us to compare the areas exposed by various crystal planes over unsupported vanadium pentoxide crystals. These studies clearly showed that samples prepared by the two temperature programmed techniques gave rise to substantially different crystal dimensions. As observed by SEM images, while the technique that involved decomposition and subsequent calcination of ammonium metavanadate produced crystallites with roughly equal side and basal planes, the samples prepared by melting and temperature-programmed recrystallization showed preferential exposure of the basal plane (010).

A comparison of the X-ray diffraction patterns provided further evidence of the preferred orientation of the V_2O_5 samples discussed above. The relative intensity of the (0k0) reflections were consistently higher in samples prepared by melting and recrystallization.

The laser Raman spectroscopy studies of the unsupported vanadia showed substantial differences between relative intensities of the bands that were observed over the two samples with preferentially exposed crystal planes. When the ratios of the intensity of 995 cm^{-1} band to those of 702 cm^{-1}, 528 cm^{-1}, and 483 cm^{-1} were compared over the two samples, these ratios were seen to be consistently higher for the sample which preferentially exposed the basal plane. Sine the 995 cm^{-1} is associated with V=O stretching vibration and the 702 cm^{-1}, 528 cm^{-1} and 483 cm^{-1} are associated with bridging oxygen sites (20), this observation is significant in showing that the sample which was prepared by melting and recrystallization had a higher concentration of V=O sites. This is in agreement with SEM observations and X-ray diffraction findings as well as the literature reports which proposed that V=O sites were concentrated over the (010) plane (6).

In the kinetic experiments maximum activity for NO conversion was exhibited by both the samples at 350°C, however the samples which preferentially exposed the (010) planes (sample M) were much more active than samples D at higher temperatures indicating that the activity is related to the number of V=O species on the surface. A careful investigation of the production rates of N_2O and N_2 for both the samples over the temperature range of 150 to 400°C indicated that the basal plane (010) contributed more towards the formation of N_2O whereas the side planes contributed to the formation of N_2.

In the past the activity of the catalyst has been correlated to the level of surface coverage (25-28) and also to the number of V=O species (2, 29, 30). The maximum activity for vanadia supported on anatase was observed at a loading level of 2.3%, which is about 1.75 times the theoretical loading required for monolayer (31). It is very likely that the dispersion may not be uniform resulting in the coexistence of submonolayer coverage with localized formation of multilayers. When the activities are based on unit weight the rutile and anatase phases are comparable. Rutile phase is more active than anatase phase when activities are based on unit surface area, which is an observation different than those reported in the literature previously (2-4).

The kinetic studies over the 2.3% vanadia/titania (anatase) over the range of temperatures 100-400°C exhibit a minimum for exit NO concentration at 300°C. The exit NH_3 concentration decreases continuously, whereas the exit N_2 concentration shows a maximum at 300°C. The N_2O production starts at 250°C and continuously increases there after, suggesting direct oxidation of NH_3 at higher temperatures.

When the role of gas phase oxygen was investigated over supported vanadia catalysts, the activity of the catalyst was seen to be a strong function of oxygen concentration below 1% level. The activity dropped to zero when oxygen concentration was below 0.4%, indicating that the presence of gas phase oxygen was essential to sustain the reaction at steady state.

Ammonia oxidation studies over the unsupported vanadia catalyst to investigate the structural specificity of vanadium pentoxide catalysts and the role of direct ammonia oxidation in product distribution and isotopic labeling studies using $^{18}O_2$ with an on-line gas chromatograph-mass spectrometer to investigate the role of lattice and gas phase oxygen are currently in progress.

Acknowledgment

This material is based upon work supported by the Environmental Protection Agency under Award No. R-815861-01-0. Partial financial support from Exxon Corporation is also gratefully acknowledged.

Disclaimer

Any opinion, findings, and conclusions or recommendations expressed in this publication are those of the authors and do not necessarily reflect the views of the EPA.

References

1. Bosch, H.; Janssen, F. *Catal. Today*, **1988**, *2*, 369.
2. Inomata, M.; Miyamoto, A.; Ui, T.; Kobayashi, K.; Murakami, Y. *Ind. Eng Chem. Prod. Res. Dev.*, **1982**, *21*, 424.
3. Wainwright, M. S.; Foster, N. R. *Catal. Rev. Sci. Eng.*, **1979**, *19(2)*, 211.
4. Vejux, A.; Courtine, P. *J. Solid State Chem.*, **1978**, *23*, 93.
5. Gasior, M.; Machej, T. *J. Catal.*, **1983**, *83*, 472.
6. Andersson, A. in *Adsorption and Catalysis on Oxide Surfaces* ; Che, M.; Bond, G.C., Eds.; Elsevier Science Publishers B. V., Amsterdam, The Netherlands, **1985**, 381-402.
7. Miyamoto, A.; Yamazaki, Y.; Inomata, M.; Murakami, Y. *J. Phys. Chem.*, **1981**, *85*, 2366.
8. Inomata, M.; Miyamoto, A.; Murakami, Y. *J. Phys. Chem.* **1981**, *85*, 2373.
9. Cai, Y.; Ozkan, U.S. *International Journal of Energy, Environment and Economy*, **1991**, *1(3)*, 229.
10. Cai, Y.; Ozkan, U.S. *Applied Catalysis*, **1991**, *78*, 241.
11. Janssen, F. J. J. G. *Kema Scientific & Technical Reports*, **1988**, *6(1)*, 1.
12. Miyamoto, A.; Yamazaki, Y.; Inomata, M.; Murakami, Y. *J. Phys. Chem.*, **1981**, *85*, 2366.
13. Wachs, I. E.; Saleh, R. Y.; Chan, S. S.; Chersich, C. C. *Applied Catalysis*, **1985**, *15*, 339.
14. Saleh, R. Y,; Wachs,I. E.; Chan, S. S.; Chersich, C. C. *J. Catal.*, **1986**, *98*, 102.
15. Went, G. T.; Oyama, S. T.; Bell, A. T. *J. Phys. Chem.*, **1990**, *94*, 4240.

16. Lopez Nieto, J. M.; Kremenic, G.; Fierro, J.L.G. *Applied Catalysis*, **1990**, *61*, 235.
17. Wachs I. E.; Chan, S. S. *Applications of Surface Science*, **1984**, *20*, 181.
18. Wokaun, A.; Schraml, M.; Baiker, A. *J. Catal.*, **1989**, *116*, 595.
19. Wachs, I. E.; Chan, S. S.; Saleh, R. Y. *J. Catal.*, **1985**, *91*, 366.
20. Beattie, I. R.; Gilson, T. R. *J. Chem. Soc. (A)*, **1969**, 2322.
21. Bond, G. C.; Tahir, S. F. *Applied Catalysis*, **1991**, *71*, 1.
22. Machej, T.; Haber, J.; Turek, A. M.; Wachs, I. E. *Applied Catalysis*, **1991**, *70*, 115.
23. Chan, S. S.; Wachs, I. E.; Murrell, L. L.; Wang, L.; Hall, W. K. *J. Phys. Chem.*, **1984**, *88*, 5831.
24. Hernandez, R. A.; Ozkan, U. S. *Ind. Eng. Chem. Res.*, **1990**, *29(7)*, 1454.
25. Kotter, M.; Lintz, H. G.; Turek, T.; Trimm, D. L. *Applied Catalysis*, **1989**, *52*, 225.
26. Rajadhyaksha, R. A.; Hausinger, G.; Zeilinger, H.; Ramstetter, A.; Schmelz, H.; Knozinger, H. *Applied Catalysis*, **1989**, *51*, 67.
27. Baiker, A.; Dollenmeier,P.; Glinski, M.; Reller, A. *Applied Catalysis*, **1987**, *35*, 351.
28. Kang, Z. C.; Bao, Q. X. *Applied Catalysis*, **1986**, *26*, 251.
29. Inomata, M.; Miyamoto, A; Murakami, Y. *Chem. Lett.,* **1978**, 799.
30. Inomata, M.; Miyamoto, A.; Murakami, Y. *J. Phys. Chem.*, **1981**, *85(16)*, 2372.
31. Haber, J.; Kozlowska, A.; Kozlowski, R. *J.Catal.*, **1986**, *102*, 52.

RECEIVED February 12, 1992

Chapter 10

Catalytic Oxidation of Trace Concentrations of Trichloroethylene over 1.5% Platinum on γ-Alumina

Yi Wang[1], Henry Shaw[1], and Robert J. Farrauto[2]

[1]Department of Chemical Engineering, Chemistry, and Environmental Science, New Jersey Institute of Technology, Newark, NJ 07102
[2]Engelhard Corporation, Menlo Park R&D, Edison, NJ 08818

The catalytic oxidation of 40 to 250 ppm trichloroethylene in air was evaluated over 1.5% $Pt/\gamma-Al_2O_3$ on a cordierite monolith with 62 cells per cm² over the temperature range 150 to 550 °C in a 2.5 cm in diameter tubular reactor at space velocities of 1,000 to 30,000 v/v/hr. Over 99.9% conversion of TCE is achieved at 30,000 v/v/hr and 550 °C with fresh catalyst. The main products from the oxidation of TCE are CO_2, HCl and Cl_2. However, trace amounts of C_2Cl_4 and CO as intermediate products are also found. The addition of 1.5% water or 0.6% methane to the feed promotes the complete conversion of chlorine to HCl and decreases the rate of production of trace byproduct perchloroethylene. Overall kinetics are first order in TCE with an activation energy of 20 kcal/mol.

Chlorinated compounds present environmental hazards in a number of different ways. They can be found as air pollutants which affect stratosphere ozone and as water pollutants which affect health because of their carcinogenicity. Chlorocarbons are very stable and are often recycled when economically feasible. However, there are times when their concentration in waste streams is very low. Under these circumstances, it is most economical to destroy them by high temperature incineration, a procedure which may produce other harmful effluents that need to be removed before discharge to the atmosphere. The purpose of our research is to devise a simpler, safer, and hopefully less expensive procedure by which trace chlorinated compounds could be destroyed. Catalytic oxidation is such an option for control of these emissions, although wet scrubbing will still be required.

The objective of this research is to evaluate the catalytic oxidation processes for a chlorinated hydrocarbon. This includes studying catalytic kinetics for the destruction of low concentrations of chlorinated hydrocarbons, effect of space velocity, feed concentration, product distribution and catalyst aging.

Since some chlorinated hydrocarbon molecules contain more chlorine atoms than hydrogen atoms, they produce Cl_2 in addition to the more desirable HCl which is easily scrubbed in an alkaline medium. An additional objective is, therefore, to find a way of providing hydrogen in order to convert all the chlorine to hydrogen chloride. We have examined the possibility of supplying the hydrogen

0097–6156/92/0495–0125$06.00/0

required with the feed in the form of either water or methane (the major component of natural gas).

There are considerable advantages in using catalytic combustion instead of high temperature incineration. Although incinerators can meet current EPA regulations for effluent emissions, they may have operational problems with combustion stability and high destruction efficiencies in the face of the flame-inhibiting properties of halogenated compounds (1). It should be noted that conventional combustion systems require back mixing (swirling), auxiliary fuel, staged burning, high temperature of operation (producing NO_x), and long average residence times (requiring large reactor volumes and consequently, high capital costs). A catalytic combustor can avoid halogen flame inhibition sufficiently to permit plug flow combustion (i.e., without significant back mixing). With combustor size requirements significantly reduced, capital and operating costs could also be brought down and on-site or mobile incineration made more economical. Futhermore, by operating at low temperature NO_x is not formed.

Experimental

Reactor System. The experiments were conducted in a laboratory-scale tubular reactor system shown in Figure 1. This system consists of a 2.5 cm inside diameter quartz tube reactor residing in a vertical three zone controlled furnace containing a known volume of platinum monolith catalyst. The middle zone was designed to maintain a flat temperature profile over the length of the catalyst monolith. Catalyst volumes of 2.5, 5.0 and 7.5 cm were used in order to vary space velocity.

A glass U-tube containing C_2HCl_3 feed in liquid form was placed in a icebath and part of the air feed was bubbled through the U-tube, becoming saturated with TCE at room temperature. The TCE containing air stream was mixed with the rest of the air before entering the reactor. The flow rates of inlet gases were measured with four calibrated Cole Parmer rotameters.

The reactor temperature was monitored with two 0.16 cm calibrated Chromel-Alumel thermocouples inserted in both sides of the quartz tube reactor and placed in the center line immediately before and after the catalyst. Since the reactor temperatures were kept below 600 °C, no corrections were made for radiation.

Analytical. The HCl concentrations were determined by reacting the gaseous effluent with a standard solution of $AgNO_3$ until all Cl^- precipitated as AgCl. The end-point was measured using a Cl^- specific ion electrode to indicate when all $AgNO_3$ was consumed. After the end-point was determined, the flask was replaced with a fresh flask containing a known volume of the $AgNO_3$ standard solution for a new incremental measurement. The increased HCl values were used as the average over the time needed to consume the standard amount of $AgNO_3$. The results were very reproducible. Drager tubes were used to measure chlorine and double check the HCl measurements.

The gases were purchased from the Liquid Carbonic Co. and used directly from cylinders. Air was of research grade purity zero air, with less than 5 ppm H_2O and less than 1 ppm hydrocarbons. The purity of TCE used is reported to be 99+ percent by the supplier.

The concentrations of feed and product compounds were measured using Hewlett Packard 5890 gas chromatographs. One has flame ionization (FID) and thermal conductivity (TCD), and the other has electron capture (ECD) and flame photometric (FPD) detectors. The CO_2, and CO were separated first on a 1/8 inch

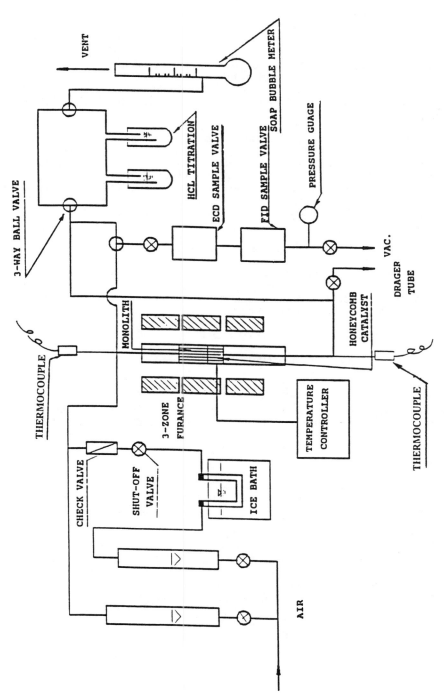

Figure 1. Flow schematic of catalytic oxidation unit.

in diameter by 6 feet long stainless steel column packed with 80/100 mesh Porapak Q and then individually hydrogenated at 350°C over a Ni-catalyst system to CH_4 before detection by flame ionization. The chlorinated hydrocarbons were separated on a 1/8 inch in diameter by 10 feet long stainless steel column packed with 80/100 mesh Chromosorb GAW and detected by electron capture.

Both carbon dioxide and carbon monoxide concentrations were calibrated with purchased standard gas mixtures. The chlorinated hydrocarbon concentrations were calibrated with the pure compounds by liquid injection. Hewlett Packard 3396A integrators were used as both recorders and integrators.

Catalysts. The catalysts used in this research were provided by Engelhard Corporation. The catalysts are 1.5% platinum deposited on γ-alumina and supported on a cordierite honeycomb with 62 cells per cm^2 (cpsc) which is equivalent to 400 cells per square inch (cpsi). To obtain high space velocity, high temperature resistant cement was used to block a number of cells, thus reducing catalyst volume. Only 25 square cells remained in the 2.54 cm diameter monolith. The length of catalyst monolith was 7.62 cm; therefore, the actual volume of modified catalyst was:

$$V_c = 7.62*(25/400)*2.54^2 = 3.0726 \text{ cm}^3$$

Different space velocities were obtained during the experiments by using different catalyst lengths. Thus, using total flow rate of 1530 cm³/min corrected to standard condition of 0 °C and one atmosphere pressure, the space velocity for 7.62cm length catalyst can be calculated as follow:

Space Velocity = total flow rate/catalyst volume

$$= (1530 \text{ cm}^3/\text{min}*60\text{min/hr})/3.0726 \text{ cm}^3$$

$$= 30,000 \text{ v/v/hr}$$

Residence times for the kinetic studies were estimated by inverting the space velocities after correcting the flow rate to the operating temperatures and pressures.

Results

Catalytic Oxidation TCE. A fresh catalyst was used to oxidize 40 ppm trichloroethylene at 30,000 v/v/hr space velocity over the temperature range 150 to 450 °C. The relationship between conversion, product distribution and temperature for the fresh catalyst is plotted in Figure 2. The product distributions, material balances for carbon and chlorine, and selectivities to CO_2 and HCl are summarized in Table I. Figure 3 shows that in the temperature range 250-330 °C, conversion increases slowly with increasing temperature, then increases rapidly to 450 °C where 95% conversion is achieved. The main products are carbon dioxide, hydrogen chloride and chlorine. Trace amounts of carbon monoxide and perchloroethylene are also produced. Almost 90% selectivity to CO_2, and only 22% selectivity to HCl was obtained. At 350 °C, the amount of CO reaches a maximum and then decreases with increasing temperature. The distribution of C_2Cl_4 follows a similar trend to CO, but peaks at 400 °C.

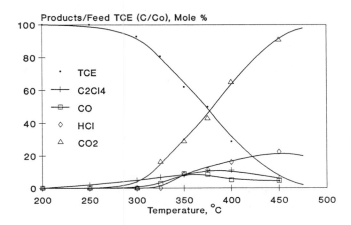

Figure 2. Product distribution from the oxidation of TCE with air over fresh 1.5% Pt/γ-Al$_2$O$_3$ on a cordierite monolith as a function of temperature. The feed concentration of TCE was 40 ppm(v) and the space velocity 30,000 v/v/hr.

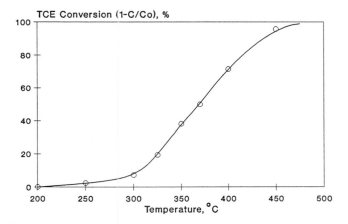

Figure 3. Conversion of TCE with air over fresh 1.5% Pt/γ-Al$_2$O$_3$ catalyst as a function of temperature. The feed concentration of TCE was 40 ppm(v) and the space velocity 30,000 v/v/hr.

Table I. Product Distribution from the Catalytic
Oxidation of TCE with Fresh Catalyst

Temp. °C	Initial TCE ppm	Products					
		TCE ppm	C_2Cl_4 ppm	CO_2 ppm	CO ppm	HCl ppm	Cl_2 ppm
150	41	41	0	0	0	0	0
250	41	41	0.8	0	0	0	0
300	41.3	38	2.0	0	0	0	0.9
325	40.5	33.1	2.8	13.8	2.4	0	5.5
350	41.1	25.4	3.3	23.7	7.3	10.6	11.7
375	42.3	20.5	4.5	35.3	6.9	14.3	16.6
400	41.7	11.8	4.6	53.2	4.1	19.7	25.8
450	40.8	1.8	2.5	74.5	3.7	27.6	39.7

Temperature °C	Material Bal.		Selectivity			Conv.
	C %	Cl %	CO_2 %	HCl %	Cl_2 %	TCE %
150	100	100	0	0	0	0
250	100	100	0	0	0	0
300	97	100	0	0	19.2	8
325	109	100	93.2	0	49.5	18.3
350	108	100	75.5	22.5	49.5	38.2
375	109	100	81.0	21.9	50.6	51.5
400	108	100	89.0	22.0	57.5	71.7
450	106	100	95.5	23.6	67.9	95.6

Notes:
      Selectivities were calculated by dividing measured product concentration by
      total possible product concentration from TCE converted
      Cl_2 effluent concentration was estimated by difference
      Space velocity = 30,000 v/v/hr
      Catalyst = 1.5%Pt/γ-Al_2O_3 on a cordierite monolith with 62 cpsc

TCE Oxidation Kinetics. The chemical kinetics of the catalytic oxidation of
trichloroethylene were determined at conversions of less than 30%. Space
velocities were varied at each temperature to allow evaluation at different
residence times. Figure 4 shows the linear relationship between log retention of
TCE and residence time. A small effect on TCE conversion due to the cordierite
monolith was found at the lowest space velocities and highest temperatures. This
effect does not exceed 2.2% conversion. The slope of each line is the first order
rate constant at that temperature. The linearity and zero intercept of the correlation
is consistent with first order kinetics. The reaction order with respect to O_2 was
independently evaluated. The concentration of O_2 in the 5 to 20% range had no
effect on the rate of oxidation of TCE. A plot of the logarithm of the rate constant
k versus the inverse of temperature for the fresh catalyst is shown in Figure 5. The

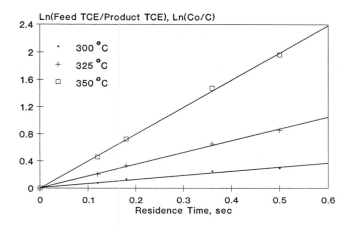

Figure 4. Determination of first order rate constant at TCE conversion of less than 30%. The feed concentration of TCE was 40 ppm(v) and the space velocity 30,000 v/v/hr.

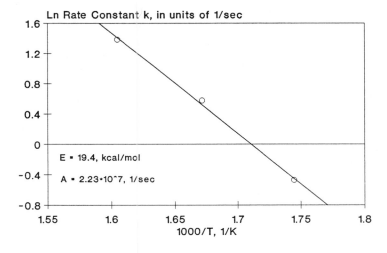

Figure 5. Arrhenius plot of first order rate constant for fresh Pt Catalyst. The feed concentration of TCE was 40 ppm(v) and the space velocity 30,000 v/v/hr.

Arrhenius parameters for this catalytic reaction are E = 19.4 kcal/mol and A = $2.23*10^7 sec^{-1}$.

Effect of Hydrogen Containing Additives. Since trichloroethylene contains only one hydrogen atom, it will dissociate at high temperatures to yield equal molar concentrations of HCl and Cl_2. Thus, for complete oxidation of TCE, one would obtain 25% HCl, 25% Cl_2 and 50% CO_2. Due to this limitation, only 22% selectivity to HCl was obtained. To improve the selectivity to HCl, either a hydrogen-rich fuel such as methane or water vapor could be added to the feed. The influence of both water and methane on product distribution for C_2HCl_3 oxidation are shown in Figures 6 and 7 and Tables II and III. With methane as the additive, the selectivity to HCl improved to more than 80% at 450 °C. Methane also inhibited perchloroethylene production. In the case of water as the additive, 100% selectivity to HCl was reached at 450 °C and water also inhibited perchloroethylene production. A comparison of conversion versus temperature for the cases of no additive, water, and methane is provided in Figure 8. The results indicate that methane can accelerate C_2HCl_3 oxidation, while water slightly inhibits this reaction.

Table II. Product Distribution from the Catalytic
Oxidation of TCE with 1.5% Water

Temperature °C	Initial TCE ppm	Products			
		TCE ppm	C_2Cl_4 ppm	CO_2 ppm	HCl ppm
150	242.0	242.0	0	0	0
265	244.3	242.0	3.2	0	0
315	240.4	235.0	4.2	17.4	12.3
365	244.7	174.2	10.7	153.5	186.3
415	242.1	93.7	11.9	303.0	461.0
465	239.8	29.8	13.5	404.5	638.7

Temperature °C	Material Bal.		Selectivity		Coversion
	C %	Cl %	HCl %	CO_2 %	TCE %
150	100	100	0	0	0
265	100	101	0	0	0
315	103	102	75.9	161	2.2
365	107	102	88.1	109	28.8
415	106	109	104	102	61.3
465	102	109	101	96	87.6

Notes:
      Space velocity = 30,000 v/v/hr
      Catalyst = 1.5%Pt/γ-Al_2O_3/monolith/62cpsc

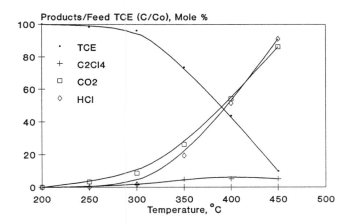

Figure 6. Effect of 1.5% water on the product distribution from TCE oxidation over fresh Pt catalyst. The TCE feed concentration was 236 ppm(v) and the space velocity was 30,000 v/v/hr.

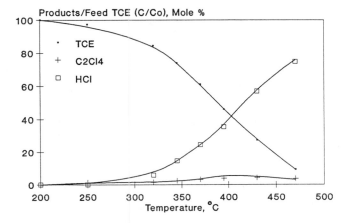

Figure 7. Effect of 0.6% methane on the product distribution from TCE oxidation over fresh Pt catalyst. The TCE feed concentration was 240 ppm(v) and the space velocity was 30,000 v/v/hr.

Table III. Product Distribution from the Catalytic
Oxidation of TCE with 0.6% Methane

Temperature °C	Initial TCE ppm	TCE ppm	C$_2$Cl$_4$ ppm	CO$_2$ ppm	HCl ppm
			Products		
150	253	253	0	0	0
265	258	248.5	0	15	0
325	247.4	208.7	4.7	54	47.3
345	252	190.8	5.1	131	72.4
370	248	148.4	6.4	237	200.2
395	255.6	112.7	8.8	472	302.2
420	261.3	80.2	8.5	1147	398
470	256.6	27.8	6.4	2407	561.2

Temperature °C	Material Balance C %	Cl %	CO$_2$ %	HCl %	Cl$_2$ %	Conversion TCE %
			Selectivity			
150	100	100	0	0	0	0
265	100	100	100	0	100	3.7
325	100	100	85.2	40.7	43.1	15.6
345	100	100	92.8	39.4	49.5	24.3
370	100	100	94.9	67.0	24.5	40.2
395	100	100	96.4	70.5	21.3	55.9
420	100	100	98.5	73.3	20.5	69.3
470	100	100	99.5	81.8	14.5	89.2

Notes:
    Cl$_2$ and CH$_4$ effluent concentration was estimated by difference
    Space velocity = 30,000 v/v/hr
    Catalyst = 1.5%Pt/γ-Al$_2$O$_3$/monolith/62cpsc

Aging Experiments.   A major concern with the catalytic oxidation of
chlorocarbons is the potential poisoning effect of chlorine or HCl reacting with the
active component, viz., Pt. Aging experiments were therefore conducted at 450
°C to evaluate if poisoning can be detected during the first 100 hours. To avoid
losing platinum because of vaporization, the maximum temperature, was limited to
450 °C. At this temperature, over 90% conversion of TCE can be achieved. A
fresh catalyst was used to oxidize 40 ppm trichloroethylene at 30,000 v/v/hr space
velocity over the temperature range 150 to 450 °C. The catalyst was cooled down
to room temperature, then reheated directly to 450 °C with a feed containing 50
ppm C$_2$HCl$_3$ at 30,000 v/v/hr space velocity and maintained for 25 hours,
followed by a cool-down to room temperature. The 25 hours aging experiments
were repeated 3 more times with the same catalyst at the same operating
conditions. Interspersed between the 25 hour aging segments, product distribution
profiles, similar to the one shown in Figure 2, were measured. Figure 9

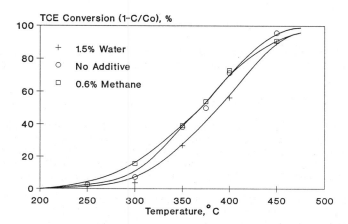

Figure 8. Effect of additives on TCE conversion over fresh Pt catalyst. TCE feed concentrations were 50 ppm(v) with no additive, 240 ppm(v) with methane, and 236 ppm(v) with water.

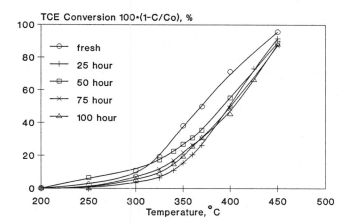

Figure 9. Effect of on-stream time on TCE conversion over the Pt catalyst. The TCE feed concentration varied between 40 and 60 ppm(v) and the space velocity was 30,000 v/v/hr.

summarizes these results by showing the variability of conversion curves with catalyst age.

Discussion

The overall stoichiometry for oxidation of trichloroethylene at 1 atm and the temperature range used in this study may be represented by the following global reaction:

$$C_2HCl_3 + 2O_2 \rightarrow 2CO_2 + HCl + Cl_2 \qquad (2)$$

However, trace amounts of CO and C_2Cl_4 were also found. Neglecting the change in total volume of gas due to the large excess of air, CO_2 concentration should be twice the amount of inlet concentration of C_2HCl_3, if all carbon atoms are oxidized to CO_2. Tables I, II, and III show that this is indeed the case. For example, we found that at 450 °C, 39 ppm of C_2HCl_3 are converted to 75 ppm CO_2.

The selectivity to HCl was calculated to be only 22% at 450 °C for TCE with no additives. This, however, is not surprising since C_2HCl_3 is a hydrogen-lean reactant containing only one hydrogen and three chlorine atoms. Thus, two atoms of chlorine are converted to Cl_2.

The formation of CO at low temperature usually occurs in homogeneous combustion followed by burnout to CO_2. Bose (2) invoked a two-step oxidation reaction of chlorinated hydrocarbons to account for this observation.

First: $C_2HCl_3 + O_2 \rightarrow 2CO + HCl + Cl_2$ \qquad (3)

Second: $CO + 1/2O_2 \rightarrow CO_2$ \qquad (4)

The concentration of CO decreases with increasing temperature because reaction 4 is favored at high temperatures.

The formation of C_2Cl_4 can be explained by platinum promoting the following free radical reactions (3):

$$C_2HCl_3 \rightarrow C_2HCl_2 + Cl \qquad (5)$$

$$C_2HCl_3 + Cl \rightarrow C_2HCl_4 \qquad (6)$$

$$C_2HCl_4 + C_2HCl_3 \rightarrow C_2HCl_5 + C_2HCl_2 \qquad (7)$$

$$C_2HCl_5 \rightarrow C_2Cl_4 + HCl \qquad (8)$$

Alternatively, one can invoke a Mars-van Krevlen mechanism for chlorine forming compounds with Pt on the metal surface. This is explored by Shaw, et al. (4).

As shown in Figure 9, only after the first 25 hours was there evidence of deactivation. Hughes (5) claimed that after an initial drop in activity, most catalysts reached a steady state. Huang and Pfefferle (6) showed that the deactivation of a platinum/γ-alumina catalyst by chlorinated hydrocarbon was reversible for hydrocarbon oxidation. Pt reforming catalyst in petroleum industry are regenerated with HCl, air and steam injection. In this research, the catalyst achieved a steady state between deactivation and regeneration since all ingredients for both effects are present in the feed (i.e., the formation of chlorine compounds may be responsible for catalyst deactivation, and HCl and water can regenerate the catalyst). This could explain why the latter three conversion curves essentially overlapped, taking normal experimental errors into account. However, detailed characterization studies would be necessary to demonstrate this effect.

The measured data for the oxidation of C_2HCl_3 in air were correlated first using the empirical power law kinetics of the form:

$$-r_{TCE} = k[C_2HCl_3]^a *[O_2]^b$$

where, $[C_2HCl_3]$ and $[O_2]$ are the concentrations for C_2HCl_3 and O_2, respectively. The linearity of the plots of the logarithm of measured retention of reactant versus reaction time shown in Figure 4 provides an excellent correlation for first order kinetics in TCE. It was also found that the rate of C_2HCl_3 oxidation is zero order with respect to oxygen at the experimental condition studied. The following rate expression correlated the experimental data for the fresh catalyst:

$$-r_{TCE} = 2.23*10^7 exp(-19440/RT)*[C_2HCl_3]$$

Activation energy, E, represents the energetics of the rate determining step on the surface of the catalyst, and the pre-exponential factor, A, represents catalyst specific characteristics. Consequently, deactivation can be correlated with changes in the A factor. The activation energies of the five different experiments were averaged, and this modified activation energy E_{av} was used in the Arrhenius equation to estimate a new A factor for each activity check. At the same temperature, the five modified A factors were calculated using the 5 measured rate constants. The plot of modified A factors versus aging times is shown in Figure 10. The modified A factors decreased after the first 25-hour aging experiment, which indicates that the catalyst is being deactivated. The mechanisms of deactivation are still unknown, and further investigation is necessary.

Since C_2HCl_3 is a hydrogen-lean reactant, to improve the selectivity to HCl, hydrogen-rich additives are required. As described above, in the catalytic oxidation of C_2HCl_3 reactions without any additive, 2/3 of the chlorine atoms are converted to Cl_2 which is an undesirable product because it is difficult to remove from the reactor exhaust gases by simple aqueous scrubbing. Consequently, it is very important to develop a method to decrease the formation of Cl_2. Based on this consideration, methane or water were successfully used and did promote the formation of HCl.

To quantify the effect of these additives, 1.5% water vapor was added to the system. Water was chosen as a hydrogen source because it decomposes at high temperature into H and OH radicals which serve as hydrogen sources, especially when a platinum catalyst is present. To the extent that chlorine atoms are present on the surface, they can react as follows:

$$Cl_2 + H \rightarrow HCl + Cl \tag{9}$$

$$Cl_2 + OH \rightarrow Cl + HOCl \tag{10}$$

$$H + Cl \rightarrow HCl \tag{11}$$

$$HOCl + H \rightarrow H_2O + Cl \tag{12}$$

In the second case, 0.6% methane was added to the gas stream. There are two reasons for choosing CH_4 as an additive. First, methane is a hydrogen rich

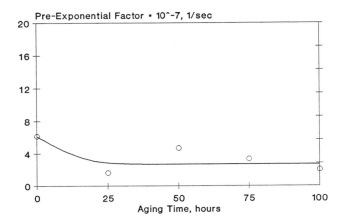

Figure 10. Variation of Arrhenius pre-exponential factor with catalyst age for an average activation energy of 20 kcal/mol. Data in the temperature range of 250 to 370 °C and conversion of less than 30% were used.

molecule which can donate hydrogen. Second, it can be used to provide heat by combustion to accelerate the reaction. The overall reaction can be written as:

$$Cl + CH_4 + O_2 \rightarrow HCl + H_2O + CO_2 \tag{13}$$

Figure 11 shows the effect of both additives on selectivity to HCl. Both water and methane dramatically increase HCl production. Figure 12 shows how both water and methane reduce yield of C_2Cl_4 as a function of temperature. Yield is defined as moles of Cl in product C_2Cl_4 per 100 moles of chlorine in the feed. The reaction mechanism probably involves removing Cl containing intermediates, and converting them to HCl.

Figure 8 shows that the addition of methane can slightly accelerate oxidation reaction of C_2HCl_3. This may simply be a consequence of the exothermicity of the oxidation of methane, increasing the surface temperature of the catalyst. The addition of water, on the other hand, slightly inhibits the reaction rate, possibly due to the endothermic water decomposition reaction.

Conclusions

- The main products from the oxidation of trichloroethylene are carbon dioxide, chlorine gas and hydrogen chloride. However, trace amounts of carbon monoxide and perchloroethylene, as intermediate products, are also found.

- Aging times of 100 hours at 450 °C had no substantial effect on product distribution. Activity reached a steady state after 25 hours of aging.

- Both water or methane, introduced as additives, increase the formation of HCl and inhibit the production of perchloroethylene and Cl_2.

- The addition of methane slightly accelerates the oxidation reaction of trichloroethylene, while water slightly inhibits the reaction.

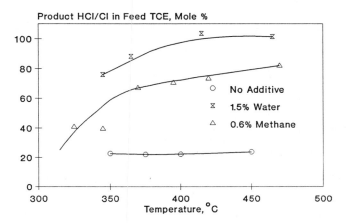

Figure 11. Effect of additives on HCl selectivity. The TCE feed concentrations were 50 ppm(v) with no additive, 240 ppm(v) with methane, and 236 ppm(v) with water.

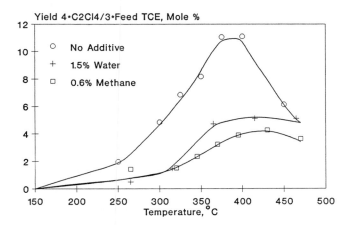

Figure 12. Yield of tetrachloroethylene as a function of additives and temperature. TCE feed concentrations were 50 ppm(v) with no additive, 240 ppm(v) with methane, and 236 ppm(v) with water.

Acknowledgments

The authors thank the NSF Industry/University Corporative Research Center for Hazardous Substance Management for support of the research. The Engelhard Corporation kindly donated all the catalysts described in this paper. The Altamira Corporation helped the researchers obtain the instrumentation to characterize the catalysts. The authors are particularly grateful to Drs. A.E. Cerkanowicz, Eric W. Stern, and Pasqueline Nyguen for enlightening discussions on the research results.

Literature Cited

1.   Bonacci, J. C.; Farrauto, R. J.; Heck. R. "Catalytic Incineration of Hazardous Waste" *Environmental Science*, **1988** *Vol. I*, Thermal Treatment. Gulf Publication.
2.   Bose, D.; Senkan, S. M. "On the Combustion of Chlorinated Hydrocarbon: Trichloroethylene" *Combustion Sci. & Tech.*, **1983** *Vol. 35*, pp. 187-202.
3.   Senkan, S. M.; Chang, W. D.; Karra, S. B. "A Detailed Mechanism for the High Temperature Oxidation of C_2HCl_3" *Combustion Sci. & Tech.*, **1986**, *Vol. 49*, pp. 107-121.
4.   Shaw, H.; Wang, Y.; Yu, T.C.; Cerkanowicz, A. E. "Catalytic Oxidation of Trichloroethylene and Methylene Chloride," I&EC Symposium on Emerging Technologies for Hazardous Waste Management, *ACS Advances in Chemistry Series*, **1992**.
5.   Hughes, R. *Deactivation of Catalysts* **1984** Academic Press Inc. London.
6.   Huang, S. L.; Pfefferle. L. D. "Methyl Chloride and Methylene Chloride Incineration in a Catalytically Stabilized Thermal Combustor" *Environ. Sci. & Tech.*, **1989** *Vol. 23, No. 9*.

RECEIVED January 28, 1992

Chapter 11

Catalytic Oxidation of Trichloroethylene over PdO Catalyst on γ-Al$_2$O$_3$

Tai-Chiang Yu[1], Henry Shaw[1], and Robert J. Farrauto[2]

[1]**Department of Chemical Engineering, Chemistry, and Environmental Science, New Jersey Institute of Technology, Newark, NJ 07102**
[2]**Engelhard Corporation, Menlo Park R&D, Edison, NJ 08818**

Trichloroethylene is completely oxidized above 550 °C to HCl, Cl$_2$ and CO$_2$ over 4% PdO on γ-Al$_2$O$_3$ as a powder or on a cordierite monolith with 62 cells per cm^2. The destruction of 200 ppm(v) TCE in air was evaluated over the temperature range 300 to 600 °C in a 2.5 cm diameter tubular reactor at space velocities of 4,000 to 24,000 v/v/hr. Only one chlorinated carbon product, C$_2$Cl$_4$, is produced at low temperatures and is destroyed above 550 °C. No significant CO formation is observed. The overall kinetics are first order in TCE. The Arrhenius activation energy is 34 kcal/mol. It was found that 1.5% water and 0.5% methane enhance selectivity to HCl and inhibit the formation of Cl$_2$ and C$_2$Cl$_4$.

Chlorinated hydrocarbons are widely used in dry cleaning, degreasing operations, and as solvents in the pharmaceutical industry, but their vapors are toxic to human beings, causing potential liver damage (*1*). Trichloroethylene (TCE) is used in this study as representative of industrial solvents. In recent years, TCE has been found in ground water and in aquifers used for potable water throughout the United States. It has become the subject of extensive governmental regulations and a target chemical in perhaps hundreds of environmental litigations (*2*). This compound, according to "Hazardous Substance Fact Sheet" (*3*), may be a cancer causing agent in humans. There may be no safe level of exposure for a carcinogen, so all contact should be reduced to the lowest possible level.

Catalytic oxidation (incineration) is an energy efficient method to destroy chlorinated hydrocarbons (*4*). Such a process involves contacting the waste gas stream with a catalyst in the presence of excess oxygen at a temperatures below about 600 °C. Chlorinated hydrocarbons are often destroyed by thermal incineration at elevated temperatures of at least 1100 °C (*5-7*). However, thermal (or conventional) incineration is very energy intensive and requires expensive materials of construction to withstand corrosion and high temperatures. Therefore, in order to reduce capital investments and fuel costs, there is a need for development of a catalyst which can oxidize chlorinated hydrocarbons at lower temperatures, while not being poisoned by chlorine. Numerous materials have been evaluated as catalysts for oxidizing hydrocarbons and chlorocarbons. These materials are usually divided into noble metals and transition metal oxides. The

0097–6156/92/0495–0141$06.00/0

activity of metal oxide catalysts for the complete oxidation of chlorinated compounds was summarized by Ramanathan, et al. (8) and Shaw, et al. (9,10).

Research in the heterogeneous catalytic oxidation of low concentrations of chlorinated hydrocarbons with air must focus on identifying highly active catalysts and obtaining benign reaction products at moderate temperatures. In general, reaction conditions are chosen that result in complete oxidation to H_2O, CO_2, and HCl (or Cl_2).

In recent years, a number of catalytic processes were developed for destroying hazardous organics in aqueous waste streams. Baker, et al. (11) developed a catalytic process which treats wastes that have insufficient heating value to incinerate, and are too toxic for biotreatment. Aqueous streams containing organics are treated with a reduced nickel catalyst at 350 to 400 °C and 20.7 to 27.6 MPa (3000 to 4000 psig), in order to convert the organics to innocuous gases.

Olfenbuttel (12) reported on new technologies for cleaning contaminated ground water. The ground water containing volatile organic compounds (VOC) can be treated by air stripping and catalytic destruction techniques. These techniques were targeted to clean up tars and oils that often coat the walls and piping of wood and agriculture waste gasifiers, and foul downstream processing equipment. Test results show that more than 90% of incoming tars and oils in the gasifier products can be destroyed in a single catalytic step.

The objective of the research presented in this paper is to enhance our understanding of catalytic oxidation processes for chlorinated hydrocarbons. In order to achieve such understanding, data must be generated on rates of catalytic destruction of low concentrations of chlorinated hydrocarbons, the effect of active metal concentration, temperature and space velocity, and hydrogen sources on chlorine and chlorinated byproducts.

Experimental

Reactor System. The experimental apparatus flow schematic is shown in Figure 1. This system consists of a quartz tube reactor (Kontes Scientific Glassware, Inc.) residing in a vertical three zone controlled furnace (Applied Test System, Inc.) containing known volumes of PdO on γ-alumina on either a honeycomb or as a powder.

A glass U-tube containing 99.9% pure TCE (Aldrich Chemical Co.) feed as a liquid is placed in an ice bath. The U-tube allows a portion of the air to bubble through the liquid before mixing with the rest of the air and feeding the mixture to the reactor. In order to increase the concentration of TCE, the proportion of air flowing into the reactant bubbler is increased. All gases were purchased from Liquid Carbonic, Inc., and were of research grade purity (99.95%). The flow rates of inlet gases are measured with two calibrated rotameters (Cole Palmer Co.).

The reactor temperatures are monitored with two calibrated Chromel-Alumel (K-type) thermocouples (Omega Engineering, Inc.) placed before and after the honeycomb catalyst. The reported temperature readings are based on average inlet and outlet temperatures. Since the reaction temperature was sufficiently low, no correction was made for radiation.

Analytical.   Feed and product chlorinated hydrocarbons are analyzed using a gas chromatograph (GC) with an electron capture detector (ECD) and a 2% SE 30 on chromosorb GAW 80/100, 1/8"x10' column. The CO, CO_2, CH_4 are analyzed using a GC with a flame ionization detector (FID), and a Porapak Q column. After

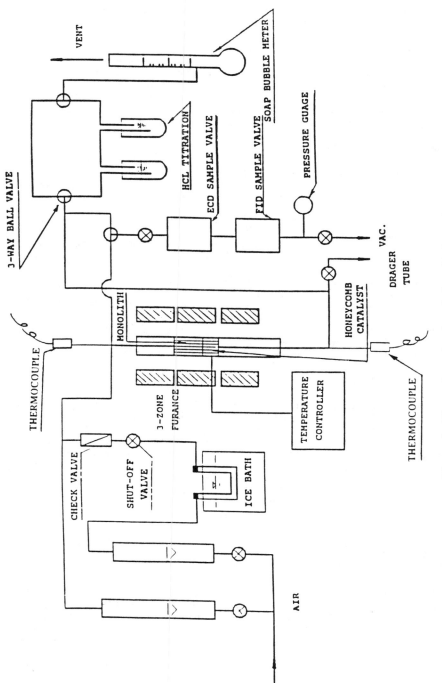

Figure 1. Flow schematic of catalytic oxidation unit.

separating the fixed gases, they are individually hydrogenated over a nickel catalyst to methane. Thus, CO and CO_2 are quantitatively measured as CH_4 which is detected by FID. Figure 2 illustrates the excellent separation obtained with the GC technique.

The concentrations of HCl were determined by absorption in a bottle of 0.01 N NaOH solution with phenolphthalein indicator, and subsequently back titrating with 0.01 N HCl solution. The concentrations of HCl were double checked with Drager gas color detector tubes. The concentrations of Cl_2 were determined by using Drager gas color detector tubes. This method of analysis was reported to be accurate within 10% according to the manufacturer.

Catalyst.   The catalysts used in this study were provided by Engelhard as either powders (50-150 mesh) with a PdO content of 0.5%, 1%, 2%, and 4% on γ-alumina, or PdO on γ-alumina supported on 62 cells per cm^2 (62 cpsc) cordierite monolith. Catalyst characterizations using temperature programmed reduction (TPR) and chemisorption (13,14) were performed using an Altamira Catalyst Characterization unit.

The activity test of powdered catalyst was run using 0.01 g catalyst, diluted with 0.1 g α-alumina. These experiments were conducted at atmospheric pressure and 340 °C. The concentration of TCE was fixed at 200 ppm and the flow rate of air was 960 cm^3/min. Figure 3 shows that conversion of TCE is proportional to available PdO content on the catalyst. Catalytic activity is therefore due to available PdO content. Hence, 4% PdO over γ-alumina on 62 cpsc cordierite is employed for most of the research reported here on catalytic oxidation of TCE.

The catalytic oxidation of TCE was conducted using 4% PdO over γ-alumina on 62 cpsc (400 cells per square inch, or 400 cpsi) cordierite in the temperature range of 300 to 550 °C and space velocities of 4,000 to 24,000 v/v/hr. The concentration of TCE was fixed at around 200 ppm and the carrier gas was dry air except when otherwise stated.

Results and Discussion

Catalytic Oxidation of TCE.   The product distribution from the catalytic oxidation of TCE at 6,000 v/v/hr as a function of temperature over the monolithic catalyst is given in Figure 4 based on data provided in Table I. The products at 550 °C are CO_2, C_2Cl_4, HCl, Cl_2 and C_2HCl_3. The conversion of TCE to the only chlorinated hydrocarbon product, C_2Cl_4, peaks at 20%. Chlorine gas is found in comparable concentrations to HCl because TCE contains more chlorine atoms than hydrogen atoms. The overall stoichiometry of the catalytic oxidation of TCE is represented by the following global reaction:

$$C_2HCl_3 + 2O_2 \rightarrow 2CO_2 + HCl + Cl_2 \tag{1}$$

It is desirable to inhibit the formation of chlorine gas and enhance the selectivity to hydrogen chloride in this system because chlorine is a toxic gas that is not as easily collected.

Methane Effect on TCE Oxidation.   The effect of methane as a hydrogen source to increase HCl formation in TCE oxidation was investigated. An additional benefit of adding CH_4 is that its oxidation is highly exothermic and increases the rate of TCE oxidation. The product distribution of TCE oxidation in the presence of 0.5% methane in air is shown in Figure 5, based on data summarized in Table II. It was found that the concentration of hydrogen chloride increases and the

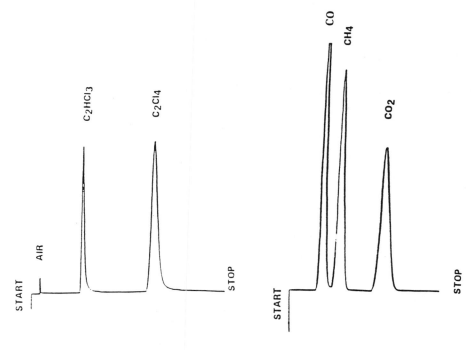

Election Capture Detection

Flame Ionization Detection
after Hydrogenation

Figure 2. Typical gas chromatograms for the chlorocarbons using electron capture detection, and CO, CH_4 and CO_2 using flame ionization detection after separation and hydrogenation with a Ni catalyst.

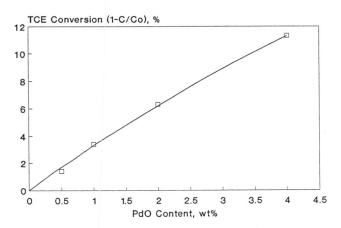

Figure 3. Activity of $PdO/\gamma-A_2O_3$ powder catalyst as a function of PdO content. These experiments were done at 340 °C with a feed of 200 ppm(v) TCE at a flow rate of 960 cm^3/min and a charge of 0.01 g of catalyst.

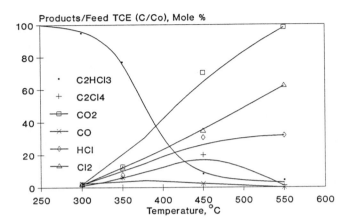

Figure 4. Product distribution from the oxidation of TCE with air over fresh 4% PdO/γ-Al$_2$O$_3$ on cordierite as a function of temperature. The feed concentration of TCE was 200 ppm(v) and the space velocity 6,000 v/v/hr.

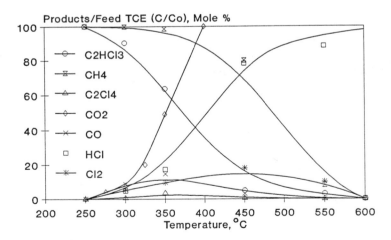

Figure 5. Effect of 0.5% CH$_4$ on the product distribution from TCE oxidation over fresh PdO catalyst. The TCE feed concentration was 200 ppm(v) and the space velocity 6,000 v/v/hr.

Table I. The Product Distribution of TCE Oxidized Over 4 % PdO/γ-Al$_2$O$_3$ on Cordierite

Temp. °C	TCE,C$_0$ Initial ppm	TCE ppm	C$_2$Cl$_4$ ppm	CO Products ppm	CO$_2$ ppm	HCl ppm	Cl$_2$ ppm	C-Balance %	Cl-Balance %
300	213	202	2.8	9.8	7.5	17.9	10	100	102
350	213	164	13.0	25.8	54.1	50.5	18	102	98.6
450	205	17.0	40.8	1.4	288	185	105	99.0	99.0
550	203	8.9	2.2	0.0	399	185	189	104	98.0

Space Velocity = 6,000 v/v/hr
Oxidant : Dry Air
Additive : None

Table II. Effect of Methane on the Product Distribution of TCE Oxidized over 4 % PdO/γ-Al$_2$O$_3$ on Cordierite

Temp. °C	TCE,C$_0$ Initial ppm	CH$_4$ Initial ppm	TCE ppm	CH$_4$ ppm	C$_2$Cl$_4$ ppm	CO Products ppm	CO$_2$ ppm	HCl ppm	Cl$_2$ ppm	H$_2$O ppm	Carbon Balance %	Chlorine Balance %
300	209	4935	188	4925	1.5	26.0	22.6	29.3	17.0	15.9	100	101
350	231	4911	147	4832	8.2	52.7	150	120	32.8	140	99.5	95.2
450	238	4878	11.3	3902	1.3	8.4	1315	559	63.8	3784	98.1	102
550	226	4901	6.6	392	0.0	0.0	4728	599	33.9	8828	95.9	101

Space Velocity = 6,000 v/v/hr
Oxidant : Air
Additive : 0.5% Methane

concentration of chlorine is reduced substantially with increasing temperature. This observation is consistent with equilibrium calculations. Formation of byproduct C_2Cl_4 is also significantly reduced. However, the rate of CO formation at low temperatures increases.

Water Effect on TCE Oxidation.   In another set of experiments, water was added to the feed in order to increase HCl formation from TCE oxidation. Water can dissociate on the oxygen precovered surface of the catalyst, providing hydrogen atoms to react with chlorinated hydrocarbons at low temperatures (*15*). Alternately, a mechanism can be written that includes surface gas phase reactions of Cl atoms with water to give HCl and OH (*16*). The product distribution from the catalytic oxidation of TCE with the addition of 1.5% water is shown in Figure 6. Activity is not affected by the addition of this relatively low quantity of water, but the rate of C_2Cl_4 byproduct formation is reduced drastically. No significant amount of CO is detected at low temperatures. The material balance data for this system are summarized in Table III. The results show that selectivity to hydrogen chloride is enhanced with water addition.

Comparison of Additives.   Table IV compares the results of catalytic oxidation with the two hydrogen bearing additives to catalytic oxidation with no additives. It is clear that the addition of hydrogen sources drives the product to HCl rather than Cl_2. This is consistent with equilibrium calculations that show a preference for HCl production with temperature when sufficient hydrogen is available.

Kinetic Study.   In these experiments, air was employed as the oxidant and was in large stoichiometric excess over TCE. In order to obtain kinetic parameters, the measured rates of oxidation of TCE in air are correlated first, using an empirical power law of the form:

$$-r_{TCE} = k_1[C_2HCl_3]^a[O_2]^b \qquad (2)$$

where, $[C_2HCl_3]$ and $[O_2]$ are the concentrations of C_2HCl_3 and O_2, respectively. The reaction order, with respect to O_2, was studied by changing the concentration of oxygen. It was found that reducing the oxygen concentration by a factor of 4 had no marked effect on the rate of oxidation of C_2HCl_3. Therefore, it was concluded that the rate of C_2HCl_3 oxidation was zero order with respect to oxygen within the experimental conditions studied. The rate reaction can be written:

$$-r_{TCE} = k_2[C_2HCl_3]^a \qquad (3)$$

where, $k_2 = k_1[O_2]$

The reaction order   with respect to C_2HCl_3 was estimated by a least square regression of equation (3). Kinetics were measured at conversions of less than 30% to insure that reaction rates are free of transport limitations. Figure 7 shows that the slope from the log-log relationship of measured reaction rate versus mean concentration of C_2HCl_3 is 0.97. Therefore, this reaction can be assumed to be first order.

In order to estimate the Arrhenius parameter for the rate constant k_2, viz., activation energy, Ea, and pre-exponetial factor, A, the method outlined by Levenspiel (*17*) or Fogler (*18*) was used. It was assumed that every channel of honeycomb catalyst is a small plug flow reactor. The equation is written as:

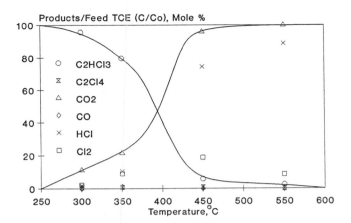

Figure 6. Effect of 1.5% water on the product distribution from TCE oxidation over fresh PdO catalyst. The TCE feed concentration was 200 ppm(v) and the space velocity 6,000 v/v/hr.

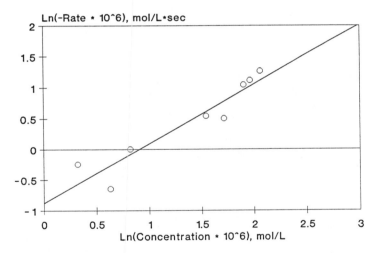

Figure 7. Determination of reaction order by obtaining the least square regression between natural logarithm of rate of TCE conversion vs. natural logarithm of mean TCE concentration. The slope, and thus, the reaction order is 0.97 at 340 °C and 6,000 v/v/hr.

$$\frac{V}{F_{Ao}} = \int_O^{X_A} \frac{dX_A}{-r_A} \tag{4}$$

where, V = bulk volume of honeycomb catalyst, cm^3.
F_{Ao} = molar feed rate, mole/sec
X_A = conversion of C_2HCl_3
$-r_A$ = rate of reaction of C_2HCl_3

Equation 4 applies in the kinetic region of the conversion curve (see Figure 4) where TCE decreases from 100 to 70% over the temperature range of 250 to 350 °C. Between 450 and 550 °C, the reaction is mass transfer limited.

Figure 8 shows plots of the logarithm of the $[C]/[C]_0$ versus residence time. A correction for a small thermal reaction component due to the monolith, which does not exceed 2.2% TCE conversion, was applied. The data are shown to fit the first order equation very well. The rate constants are obtained from the slopes of the lines. The slopes were calculated by linear regression.

Table III. Effect of Water on the Product Distribution of TCE Oxidized Over 4 % $PdO/\gamma-Al_2O_3$ on Cordierite

Temp.	TCE,C_0 Initial	TCE	C_2Cl_4	CO	CO_2 Products	HCl	Cl_2	C -Balance-	Cl -Balance-
°C	ppm	ppm	ppm	ppm	ppm	ppm	ppm	%	%
300	214	204	1.0	0.0	47.6	11.1	6.6	107	99.7
350	233	185	1.9	0.0	100	72.5	31.2	102	98.8
450	219	12.5	1.8	4.4	420	512	61.3	103	103
550	195	4.5	1.0	0.0	392	538	25.0	103	104

Space Velocity = 6,000 v/v/hr
Oxidant : Air
Additive : 1.5% Water

Table IV. Ratio of Chlorine Atoms in Products to Chlorine Atoms in Feed for the Air Oxidation of 200 ppm TCE at 550 °C and 6,000 v/v/hr

Product	No Additive %	0.5% Methane %	1.5% Water %
HCl	30.4	88.3	92.0
Cl_2	62.1	10.0	8.5
C_2Cl_4	1.4	0.0	0.7

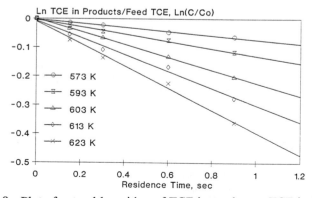

Figure 8. Plot of natural logarithm of TCE in product to TCE in feed as a function of residence time and temperature. The slope of each constant temperature line represents a rate constant.

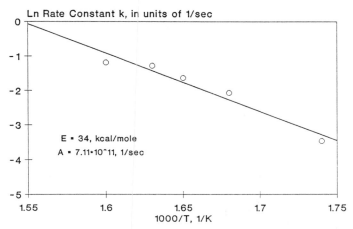

Figure 9. Arrhenius plot of first order rate constant for TCE oxidation over 4% PdO/γ-Al$_2$O$_3$ catalyst on a cordierite monolith.

According to the Arrhenius equation, the rate constant can be described as follows:

$$Lnk = LnA + (-E/R)(1/T) \tag{5}$$

where, A is the preexponential factor,
E is activation energy, kcal/mole,
R is the gas constant, 0.001987 kcal/mole-K.

Figure 9 is a plot of the logarithm of the rate constant versus 1/T. The E and A were obtained from the slope and intercept. The Arrhenius activation energy, E, is estimated as 34 kcal/mole and the pre-exponential factor, A, is $7.1*10^{11}$ sec^{-1}.

Conclusions

The catalytic oxidation of trichloroethylene over 4% PdO/γ-alumina on 62 cells per cm^2 cordierite was investigated. It is concluded that:

- The only products formed from the catalytic oxidation of TCE are HCl, Cl$_2$ and C$_2$Cl$_4$.
- Methane enhances TCE oxidation at lower temperatures, improves selectivity to HCl, and reduces production of C$_2$Cl$_4$ and Cl$_2$.
- Water improves selectivity to HCl, and reduces production of C$_2$Cl$_4$ and Cl$_2$.
- On the assumption that TCE oxidation obeys a first order rate law, an activation energy of 34 kcal/mole and pre-exponential factor is $7.1*10^{11}$ sec^{-1} are estimated.

Acknowledgments

The authors thank the NSF Industry/University Cooperative Research Center for Hazardous Substance Management for support of the research. The Engelhard Corporation kindly donated all the catalysts described in this paper. The Altamira Cooperation helped the researchers obtain the instrumentation to characterize the catalysts. The authors are particularly grateful to Drs. A.E. Cerkanowicz, Eric W. Stern, and Pasqueline Nyguen for help in the literature review and analysis of the results.

Literature Cited

1.   Bond, G. C., "Catalytic Destruction of Chlorinated Hydrocarbons", **1973**, US Patent No. 1485735.
2.   Schaumburg, F. D., "Banning Trichloroethylene: Responsible Reaction or Overkill", *Environmental Sci. Tech.*, **1990**, *Vol. 24, No. 1*, pp. 17-22.
3.   New Jersey Department of Health, "Hazardous Substance Fact Sheet; Trichloroethylene", **1986**, Trenton, NJ 08625.
4.   Lester, George R., "Catalytic Destruction of Organohalogen Compounds", **1990**, International Patent BOID 53/36, BOIJ 23/64, A62D3100, International Publication Number WO 90/13352.
5.   Bose, D.; Senkan, S. M., "On the Combustion of Chlorinated Hydrocarbons, Trichloroethylene", *Combustion Sci. & Tech.*, **1983**, *Vol. 35*, pp. 187-202.
6.   Senkan, S. M.; Chang W. D.; Karra, S. B., "A Detailed Mechanism for the High Temperature Oxidation of C_2HCl_3", *Combustion Sci. & Tech.*, **1986**, *Vol. 49*, pp. 107-121., and Senkan, S. M.; Weldon, J., "Catalytic Oxidation of CH_3Cl by Cr_2O_3", *Combustion Sci. & Tech*, **1986**, *Vol. 47*, pp. 229-237.
7.   Lee, C. C.; Talyor, P. H.; Dellinger, B., "Development of a Thermal Stability Based Ranking of Hazardous Organic Compound Incinerability", *Environ. Sci. & Tech.*, **1990**, *Vol. 24*, pp. 316-328.
8.   Ramanathan, K.; Spivey, J. J., "Catalytic Oxidation of 1,1-Dichloroethane", *Combustion Sci. & Tech.*, **1989**, *Vol. 63*, pp. 247-255.
9.   Shaw, H; Du, J.; Cerkanowicz, A. E., "The Oxidation of Methylene Chloride Over Manganese Dioxide Catalyst," Presented as paper 60a at the AIChE Pittsburgh National Meeting, **1991**.
10.   Shaw, H.; Wang, Yi; Yu, T.C.; Cerkanowicz, A. E., "Catalytic Oxidation of Trichloroethylene and Methylene Chloride," I.&E.C. Symposium on Emerging Technologies for Hazardous Waste Management, ACS Advances in Chemistry Series, **1992**.
11.   Baker, E. G. et al., "Catalytic Destruction of Hazardous Organics in Aqueous Wastes: Continuous Reactor System Experiments", *Hazardous Waste & Hazardous Materials*, **1989**, *Vol. 6, No. 1*.
12.   Olfenbuttel, R., "New Technologies for Cleaning up Contaminated Soil and Groundwater", A Battelle Special Report on Soil Remediation, **1991**.
13.   Huang, T. J.; Yu, T. C., "Effect of Calcination Atmosphere on CuO/alumina Catalyst for CO Oxidation", *Applied Catalysis*, **1989**, *Vol. 52*, pp. 157-163.
14.   Hurst, N. L. et al, "Temperature Programmed Reduction", *Cata. Rev. & Sci. Eng.*, **1982**, *Vol. 24 (2)*, pp. 233-309.
15.   Heras, J. M.; Viscido, L., "The Behavior of Water on Metal Surfaces", *Cat. Rev. & Sci. Eng.*, **1988**, *Vol. 30 (2)*, pp. 281-338.
16.   Narayanan, S.; Greene, H. L., "Deactivation by H_2S of Cr_2O_3 Emission Control Catalyst for Chlorinated VOC Destruction", Technical Report Data, for Presentation at 83rd Annual Meeting, AWMA, Pittsburgh, PA, June 24-29, **1990**.
17.   Levenspiel, O., "*Chemical Reaction Engineering*", 2nd Edition, John Wiley & Sons, Inc., NY, NY, **1972**.
18.   Fogler, H. Scott, "*Elements of Chemical Reaction Engineering*", **1986**, The Southeast Book Co., NJ.

RECEIVED March 3, 1992

Chapter 12

Thermal Decomposition of Halogenated Hydrocarbons on a Cu(111) Surface

Jong-Liang Lin and Brian E. Bent

Department of Chemistry, Columbia University, New York, NY 10027

An important step in the catalytic degradation of halogenated hydrocarbons on transition metal surfaces is the dissociation of the carbon-halogen (C-X) bond. We have investigated this process for a series of C_1 - C_3 alkyl bromides and iodides on a Cu(111) surface under ultra-high vacuum conditions. Using a variety of surface analysis techniques, we find that bromopropane and all of the alkyl iodides dissociate at temperatures below 200 K to produce adsorbed alkyl groups and halogen atoms. The alkyl groups subsequently decompose at temperatures above 200 K to evolve gas phase hydrocarbons and hydrogen; the halogen remains adsorbed on the surface up to 900 K. The activation energies for C-X bond dissociation on the surface are approximately 15% of the gas phase bond energies. A Lennard-Jones picture of dissociative adsorption is used to rationalize the observed trends in carbon-halogen bond dissociation.

Halogenated hydrocarbons have found many applications in our society as propellants, refrigerants, solvents, and synthetic reagents. These applications have also been accompanied by significant environmental impacts, both in the upper atmosphere and in ground water aquifers. Catalytic degradation of halogenated hydrocarbons is, therefore, an important current issue in environmental chemistry.

Copper as a Catalyst for Degrading Halogenated Hydrocarbons

Several observations suggest that copper might be a viable catalyst for decomposing halogenated hydrocarbons under reducing conditions. In particular, copper is well known to organic chemists as a reagent for abstracting halogen atoms from alkyl and aryl halides [1]. In addition, since copper does not form a stable bulk carbide, there is the possibility that catalyst deactivation by carbon build-up can be minimized. One might, therefore, envision using copper to decompose halogenated hydrocarbons according to the following general (nonstoichiometric) equation:

0097–6156/92/0495–0153$06.00/0

$$RX \xrightarrow{\text{Cu}} C_xH_y + H_2 + HX + CuX$$

Copper, as a catalyst, has traditionally been most used in making bonds, i.e. synthetic reactions such as methanol formation [2] and the direct synthesis of methylchlorosilanes from silicon and methyl chloride [3]. This situation reflects the inability of copper surfaces to readily dissociate ("activate") such starting materials as paraffins, olefins, and H_2. The solution phase chemistry of copper as well as the direct synthesis of methylchlorosilanes indicates that alkyl halides are an exception. Because of the relatively weak carbon-halogen (C-X) bond in alkyl chlorides, bromides, and iodides (~85, 70, and 55 kcal/mol respectively [4]), copper surfaces can activate these molecules. From an environmental standpoint, C-X bond dissociation is the key issue in the degradation of halogenated hydrocarbons, since it is the photochemical and thermal reactions of this bond that produce the environmental impact.

In this paper we report studies of the surface reactions involved in the decomposition of alkyl halides on a Cu(111) surface. We show that Cu(111) readily abstracts halogen atoms from alkyl iodides and longer chain alkyl bromides with subsequent evolution of hydrocarbon products. A Lennard-Jones picture of dissociative adsorption is used to account for the observed trends in carbon-halogen bond dissociation.

Experimental

The reactions of alkyl bromides and iodides with a Cu(111) single crystal surface were studied using an ultra-high vacuum (UHV) apparatus equipped with capabilities for surface cleaning by ion sputtering and surface analysis by Auger electron spectroscopy (AES), high resolution electron energy loss spectroscopy (HREELS), and temperature-programmed reaction studies. The experimental details will be published elsewhere [5], but several significant aspects will be reviewed here.

The Cu(111) single crystal (Cornell Materials Research Laboratory) was cleaned by cycles of sputtering at 920 K with 1 keV Ar^+ ions and annealing in UHV at 950 K. All of the alkyl halides were adsorbed onto the surface by backfilling the chamber. Exposures are given in Langmuirs (L) [1 Langmuir = 10^{-6} Torr·sec], and are uncorrected for differing ion gauge sensitivities. For the compounds used in these studies, exposures of 3-6 L are required to form a monolayer. All of the alkyl iodides and bromides except CH_3Br were used as received (Aldrich) after several freeze-pump-thaw cycles under vacuum. CH_3Br, a gas, was obtained from Matheson. Sample purities were verified in situ by mass spectroscopy. Note: Since all of these halides are possible carcinogens/mutagens, they were handled with care and vacuum pumps were properly exhausted from the laboratory.

In the temperature programmed reaction (TPR) experiments, the adsorbate-covered surface was held 1-2 mm from a 2 mm diameter skimmer to the differentially-pumped mass spectrometer and heated at 2.5 K/s. Between each TPR experiment, a clean surface was restored by briefly annealing at 980 K to remove halogens from the surface. In the HREELS studies, the spectrometer was operated at a beam energy of 3 eV and a resolution of 70 - 90 cm^{-1} full-width-at-half-maximum. All spectra were taken in the specular direction at either room temperature or 120 K after briefly annealing to the desired temperature.

The surface work function change upon adsorption or reaction was determined by measuring the change in the voltage required to "cutoff" the current to ground when an electron beam is incident on the surface. The experimental set-up for these measurements was analogous to that in reference 6, but slightly modified to suit the configuration of our apparatus. Specifically, because the crystal could not be positioned normal to the electron beam from the HREELS monochromator, the electron beam from the Auger spectrometer was used. The beam was operated at about 6 eV, and the current to ground at the crystal was on the order of 10^{-8} amperes. The experimental accuracy based on a comparison with results in the literature [7] for CO adsorbed on Cu(111) is ~ 50 meV.

Results and Interpretation

Presentation and interpretation of the results on the decomposition of the alkyl iodides and bromides will be given according to the type of experiment. The temperature programmed reaction (TPR) studies presented first establish which molecules decompose and the decomposition products. High resolution electron energy loss spectroscopy (HREELS) spectroscopy studies are then used to show that carbon-halogen bond dissociation initiates the decomposition, while work function measurements establish the dissociation temperatures.

Temperature-Programmed Reaction (TPR) Studies. Temperature-programmed reaction experiments on the alkyl bromides and iodides show that bromopropane and all of the iodides dissociate on Cu(111) at submonolayer coverage. This conclusion can be reached by monitoring either molecular desorption or the volatile hydrocarbon decomposition products as a function of exposure. We look first at the decomposition products of the alkyl iodides.

No molecular desorption is observed for the alkyl iodides at exposures less than 3 L. The major hydrocarbon decomposition products for these submonolayer exposures are shown by the temperature-programmed *reaction* (TPR) spectra in Figure 1. For iodoethane and 1-iodopropane, the major hydrocarbon product is the corresponding alkene, which is evolved at 200 - 250 K. The hydrogen atoms abstracted in forming this product recombine and desorb as H_2 at 300 - 400 K, the temperature range for recombinative hydrogen desorption on copper surfaces [8]. Some alkane is also produced at 200 - 250 K at near monolayer coverages. In contrast to the low-temperature evolution of hydrocarbon products for the C_2 and C_3 iodides, iodomethane evolves methane, ethylene, and propylene above 400 K. As shown in Figure 1, all three products have a peak temperature of 450 K, suggesting a common rate-determining step. Ethane is also produced at slightly lower temperature at near saturation coverage, but no hydrogen desorption is observed. Auger electron spectra show that iodine remains bound to Cu(111) after TPR experiments up to 900 K. All detectable carbon is removed from the surface by the hydrocarbon products discussed above. We conclude that copper surfaces dissociate the carbon-iodine bond in alkyl iodides to produce adsorbed iodine atoms and gas phase hydrogen/hydrocarbons. Further comments on the decomposition mechanisms are deferred to the discussion.

For the alkyl bromides, the extent of decomposition on Cu(111) is a strong function of chain length. This fact is evident in the temperature-programmed *desorption* (TPD) spectra of the C_1 to C_3 alkyl bromides in Figure 2. While bromomethane and bromoethane both show molecular desorption peaks for these 2 L exposures, essentially no molecular desorption is detected for bromopropane. The implication is that the C_3 bromide dissociates while the C_1 and C_2 bromides desorb molecularly intact. This observation is quantified in Figure 3 which plots the area of the molecular desorption peak as a function of exposure for these alkyl bromides. The straight lines are least squares fits to the data. The (0,0) intercept for bromomethane is consistent with 100 % molecular desorption, while the intercept at ~0.2 L for bromoethane suggests some dissociation at low exposures. Consistent with this observation, we find that ethylene is evolved at 235 K for these low bromoethane exposures. This product is analogous to ethylene evolution from iodoethane at 245 K in Figure 1. Bromopropane shows significant decomposition on Cu(111). The major decomposition products (analogous to iodopropane) are propylene and hydrogen.

Several comments on the the two peaks in the molecular desorption spectra of bromomethane and bromoethane in Figure 2 are warranted. The higher temperature

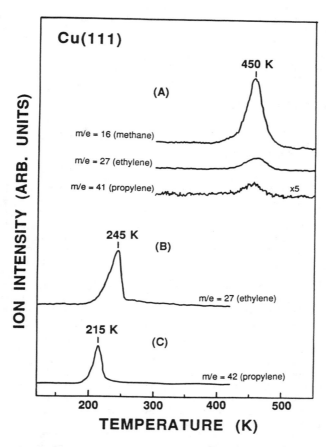

Figure 1: Temperature-programmed reaction spectra of the indicated ions after exposing Cu(111) to (A) 1 L CH$_3$I, (B) 1 L C$_2$H$_5$I, and (C) 1.5 L C$_3$H$_7$I. The hydrocarbon products responsible for each ion are noted in parentheses; propylene cracking in the mass spectrometer contributes only slightly to the m/e = 27 (ethylene) signal for CH$_3$I.

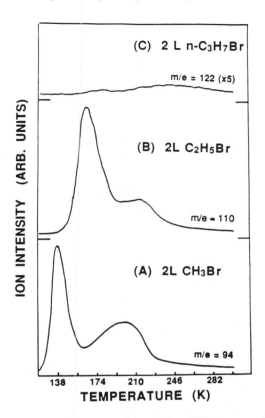

Figure 2:   Temperature-programmed desorption (TPD) spectra for the C_1 to C_3 alkyl bromides after 2 L exposures to Cu(111) at 120 K.

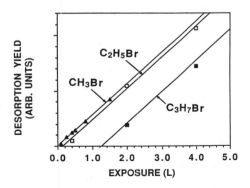

Figure 3:   Molecular desorption yield for the C_1 to C_3 alkyl bromides from Cu(111) as a function of exposure at 120 K. The points were determined by integrating TPD spectra such as those in Figure 2. The straight lines are least squares fits to the data, and yields for exposures in excess of 5 L (not shown) are included in the fit for C_3H_7Br. The arbitrary units are different for each compound.

peaks which grow in first as a function of exposure are attributed to desorption from surface defects. A similar "defect desorption state" for benzene on this crystal surface was previously observed [9]. Based on the area of the bromomethane defect peak relative to the maximum desorption yield (no multilayers of this molecule could be condensed at the adsorption temperature of 110 K), there are 10 % defects on this Cu(111) surface. For bromoethane, the smaller defect/terrace peak ratio (compared to that of bromomethane) reflects the fact that a small fraction of bromoethane decomposes at defect sites on the Cu(111) surface.

High Resolution Electron Energy Loss Spectroscopy (HREELS). High resolution electron spectroscopy studies of the bromide and iodide monolayers on Cu(111) as a function of surface temperature show that carbon-halogen (C-X) bond dissociation initiates the thermal decomposition of these compounds. C-X bond scission precedes C-H or C-C bond dissociation, since stable alkyl groups can be isolated on the surface. HREELS spectra of these alkyl groups for the C_1 to C_3 iodides are shown in Figure 4. Figure 4A is the vibrational spectrum for adsorbed methyl groups formed by dissociatively adsorbing 6 L of CH_3I on Cu(111) at 320 K. Even without assigning the surface vibrational spectrum, we can conclude that methyl groups and iodine atoms are present on the surface under these conditions based on the TPR results. The TPR results show that (1) CH_3I dissociates based on the lack of molecular desorption, (2) dissociation must occur below 300 K, since the heat of CH_3I molecular adsorption is presumably similar to that for CH_3Br which desorbs below 200 K, and (3) no C-H bonds are dissociated since recombinative hydrogen desorption was not observed at 300 - 400 K.

The surface vibrational spectrum in Figure 4A is consistent with the conclusion that $CH_3(a)$ is formed by dissociation of CH_3I. The observed peaks at 370, 1200, and 2835 cm^{-1} are attributable to the $v(Cu-CH_3)$, $\delta_s(CH_3)$, and $v_s(CH_3)$ modes respectively. (The shoulder at 270 cm^{-1} may be due to the Cu-I stretching mode.) The fact that only totally symmetric modes of C_{3v} symmetry are observed in the specular HREELS spectrum indicates that the methyl group bonds with its 3-fold axis normal to the surface. Based on the low frequency of 370 cm^{-1} for the methyl-copper stretching mode, we conclude that $CH_3(a)$ bonds in a 3-fold hollow site [10]. Note also that the observed peak frequencies are similar to those attributed to $CH_3(a)$ on Ni(111) [11] and are inconsistent with those expected for $CH_2(a)$ [12] or $CH(a)$ [13].

The two most diagnostic features for C-X bond dissociation in the $CH_3(a)$ HREELS spectrum are the absence of a C-I stretching peak at 500 - 550 cm^{-1} and the relatively low frequency for the $v_s(CH_3)$ mode at 2835 cm^{-1}. The same is true for the C_2 and C_3 iodides, as shown by the HREELS spectra for ethyl and propyl groups in Figures 4B and C respectively. Note in particular the "soft" C-H stretching frequency at ~2740 cm^{-1} in each case. Based on selective deuteration studies discussed elsewhere [14], these softened C-H stretching modes are due to the C-H bonds on the α-carbon (the one attached to the surface). The remaining modes in the spectrum are consistent with those expected for adsorbed ethyl and propyl groups, and are quite different from those for the product ethylene [15] and propylene [5]. TPD studies show that β-hydride elimination, as opposed to product desorption, is the rate-determining step in the evolution of these alkenes at 200 - 250 K [5,16].

Work Function Measurements. Measurements of the change in the surface work function after annealing monolayers of alkyl halides on Cu(111) to successively higher temperatures indicate the temperatures at which the C-X bonds dissociate. Figure 5 shows the work function change when monolayers of (A) iodomethane, (B) 1-iodopropane, and (C) 1-bromopropane are annealed to successively higher temperatures. Since the work function could not be measured while heating the surface, each of the points in Figure 5 was determined by heating the crystal at 2.5 K/s to the

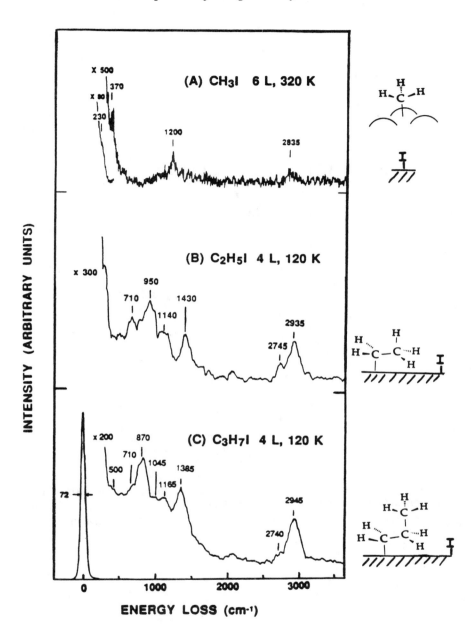

Figure 4: Specular high resolution electron energy loss spectra of Cu(111) after the indicated exposures of the C_1 to C_3 iodides. Assignment of the spectra to adsorbed alkyl groups as shown by the schematics is discussed in the text.

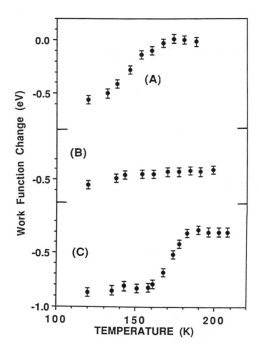

Figure 5: Work function change as a function of flashing temperature after exposing Cu(111) at 120 K to (A) 1 L CH_3I, (B) 4 L C_3H_7I, and (C) 5 L C_3H_7Br. The changes in the surface work function are due to carbon–halogen bond dissociation on the surface.

temperature indicated and then quenching to 120 K where the work function change was determined as described in the experimental section. Because the alkyl halides were not readsorbed between each of these measurements, i.e. each curve represents a single alkyl halide exposure, we cannot determine the C-X bond dissociation kinetics. The curves in Figure 5 provide an approximate indication of the work function change as a function of surface temperature.

The conclusions from the work function measurements in Figure 5 are that iodomethane dissociates at ~160 K, 1-bromopropane at ~180 K, and 1-iodopropane upon adsorption at 120 K on Cu(111). These conclusions are supported by the HREELS spectra as a function of surface temperature [5]. The reasons for these various C-X bond dissociation temperatures are discussed below.

Discussion

The results above establish that carbon-halogen bond dissociation is the initiating step in the decomposition of alkyl iodides and bromides on Cu(111). Based on recent studies for other metals, it appears that the decomposition of alkyl halides on metals is in general initiated by C-X bond dissociation to the metal surface [17-19]. In the case of Cu(111), two features of this dissociative adsorption reaction are particularly noteworthy:

(1) The activation energies for carbon-halogen bond dissociation are much less than the gas phase bond dissociation energies. Assuming first order preexponential factors of 10^{13} s^{-1} for this dissociation, the activation energies range from <7 kcal/mol for C_2H_5I and C_3H_7I to ~11 kcal/mol in C_3H_7Br. These values are approximately 15% of the gas phase bond energies.

(2) Although the gas phase bond dissociation energies for bromoethane and bromopropane are both 68 kcal/mol [4], the dissociation *yields* of these compounds on Cu(111) are dramatically different. Most of the bromopropane dissociates; most of the bromoethane desorbs.

These observations can be understood qualitatively using a 1-dimensional potential energy diagram. Such a diagram is shown in Figure 6 for the case of bromoethane and bromopropane dissociation on Cu(111). In this Lennard-Jones picture of dissociative adsorption, whether or not an adsorbate dissociates is determined by the crossing point of the molecular and dissociative adsorption potentials. Since, at this crossing point, bonds to the surface are forming at the same time the adsorbate bond is dissociating, the barrier for bond dissociation at the surface will be much less than in the gas phase. In the case of the bromoalkanes it is assumed, as shown, that the dissociative binding curve for the alkyl + bromine atom is independent of the alkyl chain length, i.e. the alkyl/surface potential is determined mainly by the Cu-carbon bond. The crossing point for the molecular/dissociation curves is therefore determined by the molecular adsorption potential. The experimental TPD results in Figure 2 show that the heat of molecular adsorption increases with alkyl chain length. The 20 K increase in the molecular desorption temperature between bromoethane and bromomethane corresponds to an increase in the heat of adsorption of 1 - 1.5 kcal/mol. This small change can have a dramatic effect on the dissociation probability if the barrier for desorption becomes larger than that for dissociation as shown in Figure 6. The C-Br dissociation temperature of 180 K for bromopropane is above the molecular desorption temperature of 160 K for bromoethane (Figure 2). The small fraction of bromoethane that desorbs from the surface defects has a molecular desorption temperature of >200 K, consistent with the partial dissociation observed at these sites.

Potential energy diagrams similar to that in Figure 6 can also be used to explain the effect of the halogen on the C-X dissociation temperature. In this case, the major effect is on the dissociation curve. For example, switching from iodide to bromide increases the dissociation potential curve far from the surface by ~ 15 kcal/mol (the difference in the gas phase bond dissociation energies). While the attractive potential between the bromine atom and the surface is also probably larger than for the iodine,

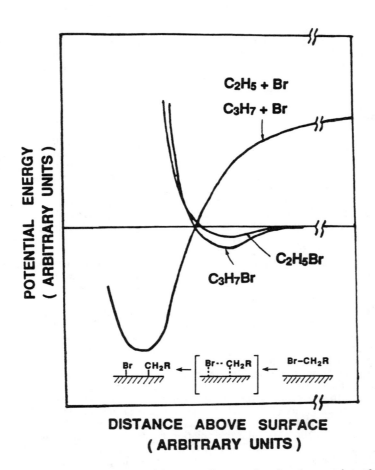

Figure 6: 1-dimensional potential energy diagram showing the crossing of the molecular and dissociative adsorption curves for C_2H_5Br and C_3H_7Br. As drawn, the crossing point for C_3H_7Br is slightly below the barrier for desorption while that for C_2H_5Br is slightly above. As discussed in the text, this type of situation can explain why the C_3 bromide dissociates and the C_2 bromide desorbs. The inserted schematic shows the type of dissociation reaction coordinate to which this diagram corresponds.

this only partially compensates for the increased gas phase bond energy. The net effect, based on the experimental observations, is that the dissociation barrier on the surface is larger for the bromides than the iodides, and in both cases the barrier is ~15% of the gas phase bond energy.

We conclude by commenting briefly on the decomposition of the alkyl groups subsequent to C-I bond scission. The mechanisms for these reactions are discussed in detail elsewhere [16,20]. In all cases, alkyl decomposition on Cu(111) is initiated by C-H bond-breaking. The difference is that C-H bond scission occurs at the β-position (1-carbon removed from the point of attachment to the surface) in the case of ethyl and propyl groups, while it necessarily occurs at the α-carbon in the case of methyl groups.

The fact that rate of β C-H bond scission is ~10^9 faster than that for α C-H bond scission [16] accounts for the very different temperatures at which the hydrocarbon products from the C_1 vs. C_2 and C_3 alkyls are evolved (Figure 1).

Acknowledgements.   Financial support from the Petroleum Research Fund administered by the American Chemical Society and from the National Science Foundation under the Presidential Young Investigator Program is gratefully acknowledged.

Literature Cited

1. (a) Fanta, P.E. *Chem. Rev.* **1964**, *64*, 613;   (b) Bacon, R.G.R.; Hill, H.A.O. *Quart. Rev.* **1965**, *19*, 95;   (c) Ebert, G.W.; Rieke, R.D. *J. Org. Chem.* **1984**, *49*, 5280.
2. See for example: Kung, H.H. *Catal. Rev. - Sci. Eng.* **1980**, *22*, 235.
3. See for example: Frank, T.C.; Falconer, J.L. *Langmuir* **1985**, *1*, 104.
4. Sanderson, R.T. *Chemical Bonds and Bond Energy*; 2nd Edition; Academic Press: NY, 1976.
5. Lin, J.-L.; Bent, B.E. manuscript in preparation.
6. Mate, C.M.; Kao, C.-T; Somorjai, G.A. *Surf. Sci.* **1988**, *206*, 145.
7. Kirstein, W.; Kruger, B.; Thieme, F. *Surf. Sci.* **1986**, *176*, 505.
8. Anger, G.; Winkler, A.; Rendulic, K.D. *Surf. Sci.* **1989**, *220*, 1.
9. Xi, M.; Bent, B.E. submitted to *Surf. Sci.*
10. Lin, J.-L.; Bent, B.E. *J. Vac. Sci. Technol.*, in press.
11. (a) Lee, M.B.;Yang, Q.Y.; Tang, S.L.; Ceyer, S.T. *J. Chem. Phys.* **1986**, *85*,1693; (b) Lee, M.B.; Yang, Q.Y.; Ceyer, S.T. *J. Chem. Phys.* **1987**, *87* , 2724.
12. See for example: George, P.M.; Avery, N.R.; Weinberg, W.H.; Tebbe, F.N. *J. Am. Chem. Soc.* **1983**, *105*, 1393.
13. See for example: Koel, B.E.; Crowell, J.E.; Bent, B.E.; Mate, C.M.; Somorjai, G.A. *J. Phys. Chem.* **1986**, *90*, 2949.
14. Lin, J.-L.; Bent, B.E. submitted to *Chem. Phys. Lett.*
15. For representative vibrational spectra of ethylene adsorbed on metal surfaces, see: Sheppard, N. *Ann. Rev. Phys. Chem.* **1988**, *39*, 589 and references therein.
16. Jenks, C.J.; Chiang, C.-M; Bent, B.E. *J. Am. Chem. Soc.* **1991**, *113*, 6308.
17. Zhou, X.-L.; White, J.M. *Surf. Sci.* **1988**, *194*, 438.
18. Zaera, F. *J. Phys. Chem.* **1990**, *94*, 8350.
19. Bent, B.E.; Nuzzo, R.G.; Zegarski, B.R.; Dubois, L.H. *J. Am. Chem. Soc.* **1991**, *113*, 1137.
20. Chiang, C.-M.; Wentzlaff, T.H.; Bent, B.E. *J. Phys. Chem.*, in press.

RECEIVED February 20, 1992

Author Index

Affiliation Index

Subject Index

Production: Paula M. Bérard
Indexing: Deborah H. Steiner
Acquisition: Anne Wilson
Cover design: Peggy Corrigan

Printed and bound by Maple Press, York, PA

Bestsellers from ACS Books

The ACS Style Guide: A Manual for Authors and Editors
Edited by Janet S. Dodd
264 pp; clothbound, ISBN 0–8412–0917–0; paperback, ISBN 0–8412–0943–X

Chemical Activities and Chemical Activities: Teacher Edition
By Christie L. Borgford and Lee R. Summerlin
330 pp; spiralbound, ISBN 0–8412–1417–4; teacher ed. ISBN 0–8412–1416–6

Chemical Demonstrations: A Sourcebook for Teachers,
Volumes 1 and 2, Second Edition
Volume 1 by Lee R. Summerlin and James L. Ealy, Jr.;
Vol. 1, 198 pp; spiralbound, ISBN 0–8412–1481–6;
Volume 2 by Lee R. Summerlin, Christie L. Borgford, and Julie B. Ealy
Vol. 2, 234 pp; spiralbound, ISBN 0–8412–1535–9

Writing the Laboratory Notebook
By Howard M. Kanare
145 pp; clothbound, ISBN 0–8412–0906–5; paperback, ISBN 0–8412–0933–2

Developing a Chemical Hygiene Plan
By Jay A. Young, Warren K. Kingsley, and George H. Wahl, Jr.
paperback, ISBN 0–8412–1876–5

Introduction to Microwave Sample Preparation: Theory and Practice
Edited by H. M. Kingston and Lois B. Jassie
263 pp; clothbound, ISBN 0–8412–1450–6

Principles of Environmental Sampling
Edited by Lawrence H. Keith
ACS Professional Reference Book; 458 pp;
clothbound; ISBN 0–8412–1173–6; paperback, ISBN 0–8412–1437–9

Biotechnology and Materials Science: Chemistry for the Future
Edited by Mary L. Good (Jacqueline K. Barton, Associate Editor)
135 pp; clothbound, ISBN 0–8412–1472–7; paperback, ISBN 0–8412–1473–5

Personal Computers for Scientists: A Byte at a Time
By Glenn I. Ouchi
276 pp; clothbound, ISBN 0–8412–1000–4; paperback, ISBN 0–8412–1001–2

Polymers in Aqueous Media: Performance Through Association
Edited by J. Edward Glass
Advances in Chemistry Series 223; 575 pp;
clothbound, ISBN 0–8412–1548–0

For further information and a free catalog of ACS books, contact:
American Chemical Society
Distribution Office, Department 225
1155 16th Street, NW, Washington, DC 20036
Telephone 800–227–5558